P

PATH

Former president of the Council for British Archaeology, Dr Francis Pryor has spent over thirty years studying our prehistory. He has excavated sites as diverse as Bronze Age farms, field systems and entire Iron Age villages. He appeared frequently on TV's *Time Team* and is the author of *The Making of the British Landscape*, and *Seahenge*, as well as *Britain BC* and *Britain AD*, both of which he adapted and presented as Channel 4 series.

FRANCIS PRYOR

Paths to the Past

Encounters with Britain's Hidden Landscapes

PENGUIN BOOKS

PENGUIN BOOKS

UK | USA | Canada | Ireland | Australia
India | New Zealand | South Africa

Penguin Books is part of the Penguin Random House group of companies
whose addresses can be found at global.penguinrandomhouse.com.

First published by Allen Lane 2018
Published in Penguin Books 2019
003

Copyright © Francis Pryor, 2018

The moral right of the author has been asserted

Set in 9.36/12.49 pt Sabon LT Std
Typeset by Jouve (UK), Milton Keynes
Printed and bound in Great Britain by Clays Ltd, Elcograf S.p.A.

A CIP catalogue record for this book is available from the British Library

ISBN: 978-0-141-98566-4

To my niece, the artist and archaeological illustrator
Chloe Watson

Contents

Introduction: Landscapes, Self and Identity   1

1   After Ages of Ice: Star Carr and the Vale of Pickering, North Yorkshire   13

2   Orkney Islands: Where Prehistory Enters Our Lives   19

3   Avebury: Much in Little   24

4   Great Orme Copper Mines: The Biggest Prehistoric Space in the World   29

5   A White Line in Time: Hadrian's Wall   33

6   Shifting Sands of Time: Seahenge, Brancaster and the Southern Wash   37

7   Arthurian Tintagel: Myths and Realities   43

8   A Haunting Place: Whitby Abbey   48

9   The Scottish Borders: The View from the North   54

10   The Boston Stump: A Fenman's Finger of Defiance   59

11   Romney Marsh: Remote, but not Entirely Forgotten   62

12   Medieval Productivity: The Open Fields of Laxton, Nottinghamshire   68

13   Ironbridge: Where It All Started   75

14   Buckler's Hard: Yesterday's Technology   80

15   The Quiet Revolution: The First Turnpike at Caxton, Cambridgeshire   84

16   A Bridge Without Sides: The Old Lower Hodder Packhorse Bridge, Lancashire   88

17   Edinburgh New Town: A Vision and Its Realization   91

18   Perfection at Rousham: Kent in Oxfordshire   95

CONTENTS

19 The Causey Arch, Co. Durham: A Long-forgotten
    Record Breaker   101

20 Birkenhead Park: The First Public 'People's Garden' 106

21 Risehill Navvy Camp: The Smell of Death,
    Above the Clouds   110

22 Lordship Rec, Tottenham: The Countryside Contained   115

23 Queensgate Shopping Centre, Peterborough: Opening
    Up City Centres   120

24 King's Cross and St Pancras: Where History Matters   125

    Notes   131
    Illustration Credits   139

Introduction

Landscapes, Self and Identity

In the autumn of 2016, my wife, Maisie, and I went to Kent for a long autumn weekend. I had endured a hernia for some time – not ideal, given that as a sheep farmer, carrying bales of straw and up-ending ewes are regular aspects of my working day – and, following a successful operation, was finding it hard to persuade my creaking body to get itself fit again.

Now you might wonder how something as intimately personal as an umbilical hernia could possibly act as inspiration for a book about the deeper lessons and underlying motives behind the formation of British rural and urban landscapes. And I must ask you to be patient for a few paragraphs. Everything will soon become clear.

I had been advised to take things easy for two months after the operation, and then to start doing more to get fit. So I heeded that advice. By early autumn I started to feel much better, and it was wonderful to be able to take walks through the garden again. We're lucky: our garden is large, covering some sixteen acres and including an orchard, an eight-acre wood and about five acres of sheltered paddocks for ewes and their youngest lambs. We had bought the land in the summer of 1992, choosing it carefully: it was quite heavy, fairly damp, and very good for establishing permanent pasture. I should also mention that the small farm that we built there, which now comprises some fifty acres, lies about six miles from the Wash, in Lincolnshire's Marshland, a north-easterly part of the much larger East Anglian Fens.

For about three years the saplings in our new wood were shorter than me. Then their roots reached the permanent water table: suddenly, they were casting shade above my head. In less than ten years

Down House, the Sandwalk: the footpath in Darwin's garden.

they were young trees. As our medieval predecessors understood so well, the soils that skirt the Wash are extraordinarily fertile: once plants become established, they take off like rockets. In planting the wood, we left clear access paths and walks, not wanting it to become a sterile plantation that nobody ever visited. The trees we chose were all British natives: oak, ash, hazel and alder. Initially, the ash, hazel and alder grew fast, but now the oaks have caught up and are over-taking them, as we originally intended. Every day for nearly a quarter of a century, barring illness, I have taken our dogs for their daily walks through the wood, along the original paths that have gradu-ally branched out, as we planted new shrubs and replacement trees, so they have now become something of a network. We have watched as wildlife has moved in: first hares and skylarks, then woodpeckers, hedgehogs and, latterly, grey squirrels, owls and buzzards; regular visi-tors include peregrine falcons and marsh harriers. The black poplars that fringe the wood have even been honoured by a pair of golden orioles, among the most spectacular of Britain's summer visitors.

In the coldest months of winter, these paths through the woods (which we have named the Siltpaths for reasons that will soon be obvious) really come into their own. This is the time of year when the farm and garden quieten down, and I do most of my writing and thinking. So, as I continued my recovery, I would set off along a path, with or without a dog, and as I walked, ideas would bubble up. Sometimes I would find myself reflecting on something I had written earlier in the morning. I still don't know how it happens, but some-how the walk helps me see things more clearly and in proportion. Ideas that had seemed brilliant a couple of hours earlier, were now plainly mediocre; other, previously trivial, hunches acquired a new importance. It seemed such a natural process, encouraged and inspired as it was by the sight of trees in the wind, glimpses of the horizon or the background chatter of angry wrens. Those post-breakfast walks allowed me to plan the day's writing in detail. I never carried a notebook, tablet or smartphone with me: if a new idea was strong enough to survive, it could wait to be jotted down when I was back indoors.

On my walks, Charles Darwin often pops into my mind. Darwin was a firm believer in the intellectual and physical benefits of a good walk. I had read, or been told, that he had done just what we had done. In 1842 Darwin acquired Down House, in Kent, which would be his home for the next half-century, and in which he wrote *On the Origin of Species*. Soon after he moved in, he laid out a series of looped paths through nearby woodland: walking and thinking paths, for keeping himself fit (he was something of a hypochondriac) and to help him work through the scientific problems with which he was currently grappling. These wooded paths, which he named the Sandwalk, played their part in shaping ideas that would change the course of Western thought. As I took my own walks, it occurred to me that the paths themselves were part of the story of how these ideas saw the light of day. The paths were part of the process of creation and discovery. They didn't just happen. Darwin had known from the outset why he had wanted them laid down.

All of which set me thinking about how details in the landscape offer us clues about the motives, the ideals, the hopes and the fears of people who, over the centuries and millennia, have engaged with it. I wanted to get away from the obvious, the usual and often mechanistic links between causes and effects: why rivers were straightened, or field boundaries curved; why railways, ports and harbours developed in certain places. These things are essential, but now I wanted to move on. I was looking for something much less specific than an economic or practical motive, or 'driver', to use a useful current term. Maybe I could find some evidence for historical frustration, pride, anger, or even for admiration, love or respect? Perhaps this was all impractical and over-ambitious? I did at least know what I had to do next: to head south, to Kent and Down House.

I have admired Darwin ever since encountering his work when studying botany and zoology in my final years at school in the early 1960s. Indeed, Darwin's writing was a major reason why I lost my tenuous hold on religion and became a lifelong secularist and rationalist. Following him, I try to see the wonders of nature, geology, chemistry, physics and astronomy for what they really are, unclouded by theological myth. And the more we discover, the more wonderful they become. The myths are no longer needed: nothing is more transcendent than knowledge.

My keenness to see Down House was more than a fan's dream. The Pryors and the Darwins were already related through the great man's wife, Emma Wedgwood. And then, five years before I was born, my uncle Mark married Darwin's great-granddaughter, Sophie Raverat. On reflection, I think it must have been Uncle Mark who first told me about Darwin's long walks at Down House. Sometimes we would visit Sophie's mother, the artist Gwen Raverat, in her Cambridge house alongside the River Cam, just upstream of Queens' College, in what was once the separate village of Newnham. I have memories of hurtling unsupervised along the corridors of the Granary, as it was called, while the grown-ups had tea. Today, appropriately enough, the house is part of Darwin College. As an undergraduate, I read Raverat's wonderful memoir *Period Piece*: few books have captured the essence of an era so vividly – or with such dry humour. The Sandwalk at Down House terrified the young Gwen at night, but by day, 'when there were grown-ups about to make it safe', she loved it, crawling 'on all fours through the undergrowth for the whole length of the wood, worshipping every leaf and bramble as I went'.[1]

Today, both Down House and Downe, the village from which Darwin's house took its name – curiously, it subsequently added an 'e' – are in the London Borough of Bromley. Despite the presence of the occasional red bus and somewhat higher levels of traffic than I am used to, the landscape was still remarkably rural: autumn tinting the trees, overgrown hedgerows and stubble fields with views across open country.

After Darwin's death in 1882, most of the family moved to Cambridge, although his widow Emma would return to Down House each summer until her death in 1896. Thereafter, the house was leased to various people and organizations, including a school, until, in 1922, it was bought from the family by a fund set up by the British Association for the Advancement of Science. It opened to the public in 1929, and is now an important visitor destination. Somehow the idea of crowds and Down House didn't appeal, so I have to confess I felt slightly apprehensive as we were driving towards it. Had it become a 'visitor attraction'?

But I shouldn't have worried: arriving at the house felt like visiting somebody's home. I half-expected someone to lean out of an upstairs window and ask if we'd both like a cup of tea. The house itself looked lived-in, not over-restored or modernized. I could imagine the Darwins' life there. Somehow, it all felt very familiar. It wasn't just that I recognized many of the pictures on the walls, which were either done by, or featured, various members of the family. It reminded me strongly of my uncle's house in Cambridge, with endless shelves of books, small cabinets of curiosities and the everyday stuff of a professional biologist's life: glass slides, tweezers, scalpels, lenses and microscopes on various benches and small tables. One of the highlights of our visit was Darwin's study, which I have to say I found very moving indeed: this was where the great man had written the books that had changed so many people's lives, including my own. These were the windows, and that was the view, he had seen when sitting at his desk, which was still there. The room had been painstakingly restored and it included furniture, pictures, books and other objects given by the family, all of which were known to have been there. But it didn't feel like a 'fossil', or something preserved in a jar. It was genuine and authentic. All in all, I found it most remarkable and I still recall it as vividly as my first view of the nave at Ely Cathedral. Both were transporting, once-in-a-lifetime experiences.[2]

The view of the garden from Darwin's study was enough to tempt me outside. Again, it didn't look over-restored: the lawn was neat, but not manicured; one or two visitors were strolling along the borders, like friends of the family after Sunday lunch. The garden seemed of its time: informal and relaxed, but nonetheless elegant. I could see from the girth of their trunks that many of the trees must have been planted in Darwin's time. Being keen vegetable growers, we went to the kitchen garden, which ran along the base of a high wall. I was surprised by the lightness and sandy feel of the soil, the kind of dry soil that needs lots of compost and manure to be dug in every season if you wanted a decent crop. No wonder Darwin was such an expert on earthworms.

Darwin liked to take long, contemplative walks every morning. This was partly for his health; he had most probably acquired a nasty infection (from the bite of a benchuga bug), while on his long and celebrated voyage aboard the *Beagle*.[3] We entered the Sandwalk off the kitchen garden, and followed it along the edge of a meadow before it looped through the Sandwalk Copse, by way of a small summerhouse.

The countryside it passes through isn't at all spectacular: there are no commanding vistas. No Gothick ruins on the skyline. Not so much as a hint of Capability Brown. You could describe it as 'gently wooded', with mature hedges, grass and the obligatory cricket pitch in the middle distance. The beauty of the scene lay precisely in this ordinariness. And I could see immediately why it was exactly what Darwin needed, when he was grappling with the intricate laws of nature. An ideal and *grounded* place for constructive thinking.

It goes – or should go – without saying that landscapes are not just about instilling pleasurable emotions. They are real, and in them real people live, and have lived, their lives; they are also constantly changing, along with the people who inhabit them. Of course today rural landscapes are under pressure as never before, because of the need to house and feed a growing population and to provide them with the services and infrastructure that modern communities expect. We have learned a few lessons from the blunders of the past, and housing estates are less frequently built on river floodplains, but increasing globalization continues to cause problems. Invasive plants and imported diseases seriously affect the countryside. If there is a good side effect of this, it's a rapidly growing awareness of the devastating impact of landscape change.

When I do something even as minor as put a fence around a field, or plant a wood, I am increasingly aware that I am altering the landscape. Twenty years ago I thought little about the impact of my actions on my surroundings; but not now. Life has become far more complex. Let's take an example. We know for a fact that many of the trees and hedges that were planted in the seventeenth, eighteenth and nineteenth centuries, as part of the landscape reforms that went with the improved farming practices of the time, were taken from cuttings from trees and shrubs that were already growing in the area. This meant that subtle regional differences in such common plants as oak, ash, hawthorn and blackthorn were maintained. But sadly, things are different today.

7

When I decided to plant an eight-acre wood, surrounded by shrubs and hedges, in the mid-1990s I faced unexpected bureaucratic problems. I could only afford to do this if I applied for a grant. It was then that I discovered that governmental and other funding agencies have to apply rules of competitive trade. So although I didn't particularly want to (normally I would have inspected the plants first), I was obliged to buy the saplings from the nursery that quoted the lowest price. I cannot be certain, but I am fairly sure that many of the trees that we planted were grown overseas. My hawthorns, for example, flower a full month earlier than those along the road outside the farm. Roughly a third of the 1,500 trees I planted were ash; as I write, many are dying of ash die-back disease, which was first recognized in England in 2012. Such problems are becoming far more frequent.

Before the industrial era, landscape change was generally incremental. But with the arrival of the railways, everything changed. I remember thinking as I looked out of a train window in the 1950s that the coal mines and slag heaps had disfigured mother nature. Nature's retribution could be terrible: the horror of the south Wales village of Aberfan in 1966 being a case in point. Disaster struck when a vast heap of waste material from a nearby coalmine slipped downhill and engulfed the village school, killing 116 children and 28 adults. Today I still feel anger at the expansion of gravel pits, which continue to destroy vast tracts of our lowland river floodplains, including thousands of known and unknown archaeological sites. Every day, hundreds of people persist in converting their front gardens to gravel-covered car parks, which support no wildlife and lead directly to rapid drainage of rainwater, which in turn contributes to the flooding of the river valleys from which the gravel was extracted. It is, to use current jargon, a no-win, or lose–lose, situation.

My point is that landscapes – whether rural or urban, large-scale or domestic – must be respected. They matter at a national level. Areas of Outstanding Natural Beauty do indeed require protection, but even these beautiful landscapes shouldn't be pickled in bureaucratic aspic; similarly, historic towns, villages and individual buildings still have to be lived in by real people and must be allowed to reflect the fashions and ideals of the age. If all our Norman churches were stripped of their later medieval additions, they would immediately lose most of their charm. In other words, landscapes must record the passage of time: they reflect human activity and have therefore acquired a life of their own. If you over-protect them, you are effectively squeezing the life out of them. They then become mere 3-D images, pictures – even caricatures – of their once-thriving selves.

The word 'landscape' was probably introduced from the Netherlands in the sixteenth century, when it referred to a painter's view.[4] While that meaning still persists, for landscape historians the word has acquired a wider significance: we regard landscape as the physical setting for human lives. Landscapes change as society itself changes, but in the process they preserve or adapt key buildings, structures or landforms. Thus Dr Beeching abolished many railway lines in the 1960s, but their cuttings, embankments and some of the bridges and stations have survived, preserved in the landscape where, over the years, they became transformed, softened, absorbed by plants and animals.[5]

Landscapes rarely yield their secrets readily. They are formed by a complex series of natural and man-made processes that have to be very carefully unravelled. Much of this work can be done by observation and a close study of maps or other documents; increasingly we are discovering answers to our questions by excavation, and surveys of the ground itself. As we all know, people – even experienced archaeologists – can misinterpret the secrets the earth yields: they can get things wrong. But when excavations and surveys reveal a coherent story that can be confirmed independently, from a number of unrelated sources, the new interpretation can be far more satisfying than the myths that preceded it.

A good example of this is our most famous prehistoric site, Stonehenge. In the Middle Ages it was seen as the work of giants and mythic heroes, but we have known for over three hundred years that this unique site is part of a complex man-made landscape that comprises many other ancient monuments. The trouble was, it was very difficult to work out how or why this landscape developed. By the 1960s we could discern the main stages in its growth, but we still lacked any understanding of the big, obvious questions. As a consequence, a number of single-cause explanations were put forward; for example, that Stonehenge was an astronomical computer. We now realize that the story of the site is immensely complex, but it does seem to make sense: at last we know why Salisbury Plain was selected; we know who built it and when they did the work. We also know where many of them came from. As a result, our knowledge of prehistoric society has been immeasurably advanced: we no longer seek 'the truth' about this unique monument, now we are starting to appreciate both its complexity and significance, and how these changed subtly over time.[6]

Most landscape histories tend to draw broad conclusions: how the Open Field System functioned in the Middle Ages, or how the railway network grew in the early 1840s. But landscapes also matter to us at a more personal level. Somehow, your relationship to your immediate landscape deepens if you are aware that the woodland you are standing in has medieval origins, or that the humps and bumps out on the village green are the remains of houses demolished after the Black Death. In urban areas, important streets are often aligned on long-lost gatehouses in the medieval town walls; railway stations and Victorian cemeteries were frequently positioned on the very edges of the built-up areas of the time. These things are not irrelevant. They link us directly to the people who created the places in which we exist and through which we move.

My own knowledge of, and approach to, the landscape has been shaped by decades in the field as a practical, digging archaeologist. So I tended to think, in the same way that I ran my excavations, that nothing was ever carried out without a clear and specific, often practical, purpose. But over the years I have discovered that often the converse was the case. Because I can appreciate how much effort went into constructing, say, Stonehenge, or a major railway tunnel, I can begin to understand the strength of the motives behind the work. And very often those motives are not just about earning a living, or the acquisition of wealth and prestige. Sometimes they revolve around a genuine wish to change the world for the better, or to create a place or space that would achieve a higher, nobler end. There is a mysticism that imbues the landscape paintings of people like Constable and Turner, which I can often detect – albeit in most instances for a fleeting, ephemeral moment – in the real world of fields, trees, lanes, Georgian squares or Victorian chapels.

Modern approaches to archaeological interpretation have taught me how simple things – an axe-haft, a sheep skull or a domestic hearth – can reveal profound truths about communities and how they lived their lives. Details such as these are what we need to make landscapes speak to us; to reveal their stories and secrets. Otherwise, they are condemned merely to look beautiful, drab or ugly, depending on the tastes and inclinations of the observer. For me, landscapes are there to inform, but also to inspire. Somehow, the landscape historian must delve deeper if he or she is ever to reveal the underlying reasons why people created great houses, new villages, or indeed railway lines. And frequently truth can be seen in the detail; for example, the way problems were solved in the course of a construction project will reveal a great deal about the motives of the people who inspired the work.

In what follows, I want to take you on a historical journey back through the British landscape, but seen from a slightly different perspective. We will examine sites and places from just after the Ice Age, some 10,000 years ago, to the most recent building at King's Cross Station, in London. It won't be a romantic or a rose-tinted view, nor will it be a search for pragmatic explanations. Instead, I want to find fresh angles and new clues: anything that will shed light on the motivations of the people who made the buildings, roads, fields and farms that together comprise our rich, varied and shared landscape.

Finally, I must confess to an ulterior motive for writing this book. For me, earth, trees, smells, insects, noises, even bricks, stones and mortar can be surprisingly direct links to times and realities that were very different from our own. Evidence from the landscape clearly shows us that, in common with most nations, Britain has always been a widely diverse place and that its distinctive regional differences evolved over many millennia. Local, regional and national identities have been greatly affected by migrants and ideas from overseas, which have had an enduring influence on our land that is still evident to this day. Other factors can be less easy to pin down. London's dominance of England, for example, can partly be explained by the layout of the road system, which radiates out from it. Most of these roads were first built by the Romans, nearly 2,000 years ago. The pattern was then cemented by the coming of the railways in the mid-nineteenth century, which essentially mirrored the road layout.

The more one understands about the British landscape, the more one appreciates the good things of the British national character, especially its tolerance and humour. Over the years I have met many new friends and colleagues, often from very different backgrounds, and the one thing we all share in common is our love of landscape history. We all have many tales to tell. So why did I write this book? That's simple: I hope the pages that follow will be your personal paths into the story of the British landscape. Then, when you have finished, you must lay the book down and venture outside. Come and join us, out there, in the real landscape. I can guarantee you won't regret it.

I

After Ages of Ice

Star Carr and the Vale of Pickering,
North Yorkshire

In the late autumn of 1967 I found myself heading north along the A1, driving flat-out in an aged Ford Popular that could just do 50 mph. Many hours after leaving Cambridge, we arrived at our destination: a chilly, flat and foggy field in what seemed like the middle of nowhere. My two friends were wildly enthusiastic and headed down the cart track as if they were approaching Valhalla. I trudged along behind them, head down, trying not to feel too sorry for myself. I have to confess, my first visit to the iconic site of Star Carr was less than memorable.

The Vale of Pickering is a wide, steep-sided valley that was sculpted by Ice Age glaciers. Flat and low-lying, it drains the higher land that surrounds it: the North York Moors to the north, the Wolds to the south-east and the Howardian Hills to the south-west. The natural drainage pattern meant that the rivers that flowed through it – principally the Derwent – would have emptied directly into the low-lying plain that would shortly become the North Sea. However, when the Ice Age climate warmed, one of the glaciers moving through the Vale began to retreat. Glaciers are famous for leaving distinctive U-shaped valleys in their wake, formed by the immensely powerful, plane-like action of slow-moving ice. Flowing glaciers scrape and bulldoze their way downhill, pushing a huge bank of rocks and earth before them. When the glaciers then stop or retreat, this bank is left stranded and can form a natural dam, known as a terminal moraine. The terminal moraine at the eastern end of the Vale still blocks all drainage into the North Sea, just south of Scarborough, with the blocked rivers forming a lake on the western side of the moraine, known to archaeologists and geologists as Lake Flixton. Today the lake has been completely drained and the land is given over to agriculture. Star Carr, the ancient settlement site the three of us visited all those years ago, lay on the shore of the onetime glacial lake.

I described Star Carr as iconic, which indeed it is, because it has revealed some of the earliest evidence for human settlement in Britain after the Ice Ages. The site was discovered before the war by local enthusiasts who had spotted flint tools lying on the surface, but it was not excavated until the early 1950s. Its archaeological importance is immense because of its location on ground that is still partially waterlogged, despite recent drainage. Waterlogging has meant that organic materials, such as wood, leaves, seeds and even pollen grains, have been preserved in the peaty muds that once formed the bed and foreshore of the glacial lake. Analysis of the pollen preserved in these peats has allowed botanists to reconstruct the changing environment at the time the site was occupied.

As a student, three lecturers inspired me. All were hands-on excavators. The best-known of them, Graham Clark, I regarded as something of a God. Admittedly, he was quite a dry and scholarly sort of God, but then deities don't always have to thunder, point doom-laden fingers, or sport imposing beards.

Clark's most famous excavation, done in the early 1950s, was Star Carr. I remember Professor Clark would frequently repeat that 'excavation without publication is simply destruction'. For me, it still remains the ultimate archaeological sin. Living up to his words, Clark ensured that his site was promptly and thoroughly reported and it illustrates very well how perceptions, interpretations and knowledge can change through time, with even the most rigorous, disciplined, science-based research.[1]

I've often wondered at what point did Clark and his team of specialists from Cambridge begin to realize that they were making what were to prove internationally important discoveries. I certainly have started to dig sites which had splendid potential, but then turned out to be ordinary, even disappointing. But was there something different about Star Carr? First of all, its location is superb and although my first trip there was less than auspicious, many subsequent visits have always proved exciting. As you head off the road and out into the peaty fields you get a distinct feeling of descending into a vanished lake. The shoreline of the glacial lake is hidden in summer when the crops are growing, but after ploughing in the autumn it is easily discerned. The landscape changes are subtle but clear, which is why Clark knew exactly where to dig. Had I been in his shoes, I'm sure I would have made precisely the same decision. My own feeling is that the whole team would have known they were on to a winner from the very beginning, which may help to explain why the project generated such excitement at the time.

The site's setting in the landscape may have been evident, but its archaeological significance has become a matter for hot debate. These are not dry academic discussions about recondite details. Rather, it's a question of whether we're talking about a tiny population of wandering hunters living in temporary camps, or was something altogether more substantial taking place? There have since been a number of important new research projects in the Vale of Pickering and we now understand very much more about the nature of its landscape some 10,000 years ago. But how were people earning a living there? What was the population? These were the basic questions that needed to be answered, with new information from the ground. It has not proved to be an easy quest and there have been many false starts and dead-ends, but at last I think we are now approaching an answer – and if you had suggested that to me ten years ago, I would probably have thought you were joking.

As was the practice at the time, Clark opened relatively small, hand-cut trenches. From them, he and his team established that Star Carr was extremely ancient and could reliably be dated to the first few centuries after the Ice Age. Today, we know that the North European climate warmed by some 10°C around 9600 BC, and it did this very rapidly, maybe over just fifty years. Then the graph naturally levelled off.[2] The period between the return of people to Britain after the Ice Age and the arrival of farming around 4000 BC is known as the Mesolithic, or Middle Stone Age. The Mesolithic settlement at Star Carr is currently dated to around 9000 BC, close to the start of the period.

Clark believed they had recovered the remains of a temporary hunting camp along the edges of the old glacial Lake Flixton, whose boundaries have since been mapped out in some detail. The Star

Carr 'camp' was on the gently sloping southern shore of a freely draining sandy promontory on the western side of the lake. Using botanical and other evidence, Clark and his collaborators concluded that the camp was probably occupied by bands of hunters, on and off, during the wetter months of winter. In summertime they would follow their prey up on to the higher land that skirts the Vale of Pickering. Clark's finds included many worked flints, together with shaped bone or antler tools and hunting spears. But the finds that brought the site to national prominence were a pair of antler headdresses, known to archaeologists as frontlets. These consisted of the uppermost skull of a red deer stag, with the antlers still in place. The rough bits on the underside of the skull had been filed or scraped off and two holes had been bored through – presumably for hide thongs to tie the headdress on to the wearer's head. Most prehistorians now agree that these frontlets were probably worn by shamans during rituals to do with hunting.

After Clark's time, research into the area continued. Some twenty-five Mesolithic sites have so far been detected around Lake Flixton, and they have revealed some of the oldest wooden artefacts in the world.[3] The most recent research project, an excavation at the edge of Clark's old site at Star Carr, has made some truly astonishing discoveries, including copious evidence for woodworking, a substantial artificial timber platform along the lakeside, and a settlement of some five acres. This settlement, of perhaps twenty-five houses, includes hundreds of thousands of worked flints and the foundations of what is now Britain's earliest house, fashioned from posts and brushwood, probably with earthen or hide walls and a roof thatched with the reeds that grew plentifully along the lakeside. Its earth floor, packed with hundreds of rejected flints, may well have been carpeted with reeds; evidence remains of a small fire, though cooking would probably have happened outside the building – to avoid accidents.

So we have come a very long way from Clark's seasonally occupied temporary hunting camp. The latest excavation suggests that Star Carr, far from being a seasonally occupied hunting camp, was lived in all year round. Our understanding of Britain's earliest-known settlement site has been transformed, and we can only assume that Star Carr wasn't unique: many of the sites around Lake Flixton

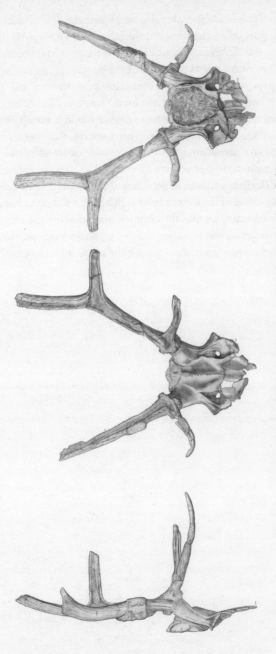

Mesolithic frontlet headdress, 9000 BC, Star Carr, North Yorkshire.

would have been of comparable size and longevity. It is also hard not to think that the landscape around the lake would have been divided up between the different villages, each with its own territory. What we witness in Star Carr is the start of the parcelling up of the land-scape, a process that would eventually give rise to the parish boundaries and field systems of our own time.

The need to have a home-base in which to raise a family is a part of being human, and the people who lived in those early round houses at Star Carr, at the end of the Ice Age, were identical to us, and not just physically. If you could have spoken their language and worn their clothes, you would soon have felt at home, sitting around their family fire, while supper gently bubbled on a spit. Soon you would be joining in, teasing the children and stroking the dogs. At a very basic level, certain places can be transforming: merging the past – even remotest prehistory – with the present.

2

Orkney Islands

Where Prehistory Enters Our Lives

The first time I came close to visiting the Orkney Islands was in the spring of 1961. I was sixteen and aboard a North Sea trawler, battling its way through the high seas of the Pentland Firth, the narrow strait which separates Orkney from the mainland, and where the currents of the Atlantic meet those of the North Sea. We had set out from Grimsby and were sailing towards the then disputed fishing grounds off the north-east coast of Iceland, during what was later to be known as the First Cod War. I had no idea of the dramas that awaited us inside the Arctic Circle. My only concerns were in the here-and-now. When I wasn't suffering bouts of violent sea-sickness, I was terrified about passing safely through the notorious Pentland Firth. Storms had been forecast, and they had arrived.

There are about seventy islands in the Orkney group that lies off the coast of Caithness, at the north-eastern tip of the Scottish mainland. From the 1500s the islands were served by a ferry, which was established by a Dutchman, Jan de Groot, who is commemorated in the name of the little port he set up: John O'Groats. Today the main ferry to the Orkney port of Stromness leaves from Scrabster, near Thurso, a few miles to the west. I would always recommend travelling to Orkney by sea, as the approach to Stromness, past the mountainous island of Hoy, with its dramatic rocky stack, the Old Man of Hoy, is spectacular.

The two principal towns of Orkney, Stromness and the capital, Kirkwall, with its superb medieval cathedral of St Magnus, are both on the largest island, Mainland. Mainland forms the northern side of the sheltered anchorage of Scapa Flow, which played an important part in the naval campaigns of both world wars. The southern

Orkneys are bounded by Hoy and the hills of South Ronaldsay. The group's northernmost limits are marked by the lower lying islands of Westray, Papa Westray and North Ronaldsay. The island of Mainland, which lies at the centre of the Orkneys, is more hilly than mountainous; it has numerous lochs and includes some of the most important prehistoric sites in northern Europe.

Orkney is an accident of history, or prehistory, to be more precise. Now bare and open, the Orkney Islands were originally covered with shrubs and trees in all but the most exposed places. Then, when the first farmers became established, after about 3500 BC, the tree cover was quite rapidly felled, probably for firewood and to clear grazing. The fierce Atlantic gales prevented young saplings from ever becoming re-established.[1] The result is the largely treeless landscape that greets the modern visitor. After this, though, why didn't people throw up their hands in despair and just move somewhere else?

The answer is the Old Red Sandstone that occurs widely across Orkney at, or just below, the surface. It is a superb natural building material: easy to quarry and even simpler to shape, it is strong enough to resist gales or accidental damage by humans and animals. As far as fuel was concerned, there were plentiful supplies of dried peat, as well as the driftwood that occurs commonly around the coasts of the Scottish Isles, some of it ultimately deriving from North America via the Gulf Stream.

Orkney possesses some of the finest prehistoric tombs and shrines in Europe. Sites like the great chambered tomb at Maeshowe, the henges and standing stones of the Ring of Brodgar, or the Stones of Stenness are extraordinary structures and have rightly acquired an international reputation. The great grass-covered mound that covers Maeshowe can be seen from great distances across lakes and fields, whereas the two stone rings that are part of the same huge landscape are very different. The Stones of Stenness are tall, sharp and in a tight group; the Ring of Brodgar is much larger, its stones evenly spaced around the edge of a huge ditched circle. Another superbly preserved complex of shrines and temples has recently been revealed at the Ness of Brodgar, the narrow spit of land that separates the lochs of Harray and Stenness, on Mainland Orkney. Here the shrines are more densely packed together, and seem to have been miraculously preserved beneath banks of sand and soil. Many of these sites were placed in landscapes on the edge of, or just beyond, land that was suitable for farming or settlement. Sometimes it was too rocky, high, marshy or tidal. These were liminal locations, places believed to be gateways to other, spiritual realms, just beyond the threshold of our own familiar world.

Because the prehistoric structures were always built from stone rather than more perishable wood, their preservation is most remarkable. This applies to ordinary, domestic buildings too. Some five miles to the north-west of Maeshowe, and overlooking the Bay of Skaill, with the open expanses of the Atlantic beyond, is the Neolithic settlement of Skara Brae, which boasts what are arguably the best preserved prehistoric houses anywhere in Europe. They appear to be nested together against the Atlantic gales, with walls still surviving to shoulder height. Even the furniture – all in carefully shaped stone – survives, with niches in walls, central hearths, even box-beds for bracken mattresses and remarkable shelved dressers.

In 2003 an unusual stone slab was discovered by local farmers on Orkney on the narrow spit of the Ness of Brodgar. The farmers realized it was probably prehistoric and potentially important, but nobody could have imagined how remarkable the site that it revealed would soon prove to be. The archaeologist who visited the site that day was Nick Card, who has subsequently directed an extraordinary series of research excavations that are still producing results that even the most hardened of prehistorians (including myself) are finding difficult to digest.[2] The wealth, complexity and preservation of the religious buildings revealed within the deep layers at the Ness of Brodgar are almost beyond belief. I first stood inside one of the buildings on a sunny, and almost windless, July morning in 2013. The site was empty because the archaeologists had just gone for their tea break, but the stone walls were so sharp, crisp and fresh, it seemed as if it was the Neolithic masons who had just left, and would return any minute to answer the questions that were filling my head. In reality, of course, their last break from work had been taken some four and a half millennia ago. The latest radiocarbon dates suggest that the shrines and temples at the Ness of Brodgar were in use for about a thousand years, from 3200–2300 BC.

I strongly suspect that the shrines and monuments that even today crowd the Orcadian landscape were originally visited by hundreds, maybe even thousands of people. Everything about the constant building and modification of sites like the Ness of Brodgar suggests they were well and frequently used. Access was normally quite good and individual shrines were enlarged and modified frequently. It is hard to avoid the conclusion that while the landscape they inhabited might have been liminal, it was far from remote or deserted.

Much of the evidence suggests that this ritual landscape may have been a destination of pilgrims, but it was also a place where people lived, fished and farmed. The teeming waters were, I imagine, very probably what attracted people there in the first place. But we should be careful of assuming that people were attracted to these places merely because they were spectacular. As we have seen, the Ness of Brodgar separates two lochs. But there was more to it than that. The Loch of Harray would have been freshwater, whereas Stenness was always open to the sea, and consequently salty. The two lochs were joined at the Bridge of Brodgar (not far from the Ness) but saltwater from the Loch of Stenness never penetrates in sufficient quantity to affect the freshwater plants of the Loch of Harray.[3] To ancient communities, this close juxtaposition of fresh and saltwater would have been remarkable and seemingly inexplicable. Fresh and saltwater lochs provide very different resources in terms of fish and plant life. A reliable supply of freshwater was essential to communities surrounded on all sides by the sea. Buildings, such as the houses at Skara Brae, even featured stone-lined water supplies that anticipated modern pipes and drains.[4]

Of course we will never be sure about what fresh or salt waters signified in the distant past, but we can be certain that it would not have been simple. There would have been realms of symbolism attached, whose richness and complexity we must never underestimate. The more we understand about the details of life in prehistory, the more we realize that it was just as intellectually complex as ours today. People did indeed inhabit the real world, but they must always have been aware of other, more important realms, just over the horizon. This is apparent in the many parallels between the religious buildings of the Ness of Brodgar and the domestic houses at

Skara Brae. You can even find echoes of domestic dwellings in the structure of tombs such as Maeshowe, or earthworks that accompany the Stones of Stenness. This raises questions about the separation of religious and secular life that are highly relevant today. At a very profound level, the ancient sites of Orkney can enter our lives – and change them.

3

Avebury

Much in Little

There are landscapes that demand stillness, where it's enough to simply stand and absorb the past all around you. Do these places have some other-worldly power, some mystical vibration? I think not, although I know many will differ from that view. I'm a rationalist and I think it somehow demeans places – and indeed people – to ascribe to them supernatural forces. The magic, I believe, comes from acknowledging and understanding the depth of human engagement with such places over time. In this, the stone circles and henges of Avebury always draw me back.

Stone and timber circles are mostly found in the British Isles, but they also occur in western France and in parts of Scandinavia. Henges, which get their name from Stonehenge, are a more specialized form of circle that developed in the British Isles. They consist of a circular ditch with an external bank, which is broached by at least one entranceway, although some monuments may have two, or even four. The placing of the bank outside the ditch contrasts with the construction of Iron Age hillforts and strongly suggests that henges were never defensive. More likely, the ditch marks the edge of a sacred area and congregations could have viewed the ceremonies taking place within the stone circles from the external bank. Most henges were built in the late Neolithic and Early Bronze Age – roughly between 3000 and 1500 BC.

Avebury is the largest henge of all and is also one of the best preserved. Like most of the larger henges, including Stonehenge and those we have just visited in Orkney (the Stones of Stenness and the Ring of Brodgar), Avebury sits at the centre of a complex ceremonial and religious, or 'ritual', landscape. As is usually the case, henges

form the latest and last stage in the evolution of these ritual landscapes, which mostly go out of use shortly after 1500 BC. But what
makes the place so relevant, for me, is the fact that it is still part of
the community. Local people don't treat it as an ancient, remote
relic. It is part of their world, with a small hamlet, including a pub,
at its centre. Despite the presence of the village, the features of the
henge are still remarkably well preserved and have clearly been
treated with respect through the ages. Many of the standing stones
are still intact. The site has never been extensively ploughed, nor
quarried, and its encircling ditch and bank are still spectacular.

Avebury sits in the rolling chalk hills of the Marlborough Downs,
some ten miles south of Swindon. It is surrounded by prehistoric
sites, such as the slightly older West Kennet Long Barrow and the
much earlier ceremonial site, in fact a precursor of henges, the so-
called Causewayed Enclosure, on Windmill Hill. Today Avebury is
shaded by tall beech trees that can be seen from some distance across
the open downland fields. But you know you have arrived somewhere unique as soon as the road passes through the outer ditch.

That massive ditch was dug out by people using antler picks, bone
shovels and baskets to carry the chalk blocks up to the surface, to
form the bulk of the great bank that runs around the outside of the
ditch. The labour involved was prodigious, but – contrary to what
some would have you believe – these stones weren't raised by the com-
mand of a powerful leader. Rather, the impetus came from the oldest
and most important element in any human community: the family.
Recent research has suggested that many monuments of the early
third millennium BC were built by groups of people who worked in
discrete groups. It misses the point to think of this as 'gang labour' in
the modern sense of the word, for there was nothing regimented or
forced about it. People carried out their tasks because they wanted to;
it was part of being a member of a clan or family. Working at Avebury
might well have marked an important stage in a person's life: maybe
a young man's transition from adolescence to adulthood. It would
have been a rite of passage.

So while Avebury is a massive, majestic place, it was conceived
and built at a very human scale. Certainly, major festivals would
have taken place there 5,000 years ago, but the great outer stone

circle was never constructed to dwarf, or to over-awe – unlike, for instance, the soaring spire of nearby Salisbury Cathedral, which seems to transcend mere humanity. Avebury remains firmly rooted in the prehistoric here-and-now.

I have a soft spot for Avebury for other reasons, too. I recall the tiny figure of Isobel Smith, late the curator of Avebury Museum, standing in the museum's doorway, very shy and reserved but nearly always smiling. It was Isobel, who had been deeply involved with the famous pre-war excavations at Avebury, who guided my journey into the Neolithic when I began my own work in the early 1970s.[1] Isobel, for me, was Avebury: modest and warm-hearted, but enormously helpful. Where Stonehenge and the great monuments of Orkney dominate in their treeless landscapes, Avebury is very different; it is slow to reveal its many secrets. You have to walk, observe and search to discover.

A mile and a half north-west of Avebury village – an easy walk, even after a good lunch at the pub – is Windmill Hill. You can find windmill hills everywhere in Britain, for obvious reasons. This one, however, is noteworthy not for its absent windmill, but for an extraordinary set of earthworks positioned slightly off the crown of the hill itself. These were first dug a very long time ago: around 3700 BC, to be more precise. Windmill Hill is not the only site of its type and age to be placed slightly off-centre – whether on a hill, or in a bend in a river – and many people, myself included, have suggested that this indicates that the place itself was considered sacred before the site was put there. People did not want their new communal shrine to dominate such a hallowed spot.

Windmill Hill consists of a strange arrangement of ditches surrounding an open space. The ditches are dug in short lengths, separated by undug 'causeways', which give this class of site its archaeological name: causewayed enclosures. Sadly, many causewayed enclosures in lowland Britain were positioned near the settlements they served, in river valley floodplains, and have since been damaged or destroyed by gravel quarries. They are, however, truly fascinating monuments and the well-preserved Windmill Hill, in its rural setting, is one of the most atmospheric. Essentially, these enclosures were probably tribal meeting grounds, where the separated communities of a region came together to exchange goods, barter for new livestock and maybe even to meet new wives and husbands.

For me, this remote part of the Avebury complex holds a special significance. In Britain, larger settlements started to become permanent fixtures in the landscape from around 4000 BC, with the arrival of farming. Very often, the new settlements were positioned close by one of the many causewayed enclosures that were just starting to be built. The enclosures were often positioned on land that wasn't suitable for arable farming, and we assume that the people who built and used them lived in the nearby settlements. But why on earth were the ditches dug in short segments, separated by causeways? The answer, which has important implications for the construction of Avebury and other henges over half a millennium later, lies in the organization of the workforce.

I have suggested that each of those many short lengths of ditch that you can still see on Windmill Hill represents a single, long-forgotten, Neolithic clan or family. This was the gang or group of people who dug the ditch segment, then filled it with religious offerings. I have excavated a similar site, which revealed copious evidence for family life: pots, flint tools and millstones for grinding corn, often carefully arranged in the ground in neat, discrete heaps, or offerings. I suspect each offering told its own story: marriage, maybe, or childbirth. So when I look at those little ditches at Windmill Hill my mind conjures up images of long-gone family gatherings: crying babies, the squeals of young children, grumpy granddads and doting parents. Those lengths of ditch may have been the focus for family gatherings, but they also provided the social cement that kept communities together. In many ways I see them as the equivalents of the family pews that were such a feature of Victorian and older rural churches.

All ancient societies attached great importance to life's rites of passage, the ultimate, of course, being death. If Avebury is at the centre of its ritual landscape, with Windmill Hill at its north-eastern edge, the West Kennet Long Barrow is on its southern edge, just over a mile from Avebury.[2] Like Windmill Hill, this is a monument to family life, but here the evidence survives in the stone-lined chambers of a large communal tomb, where bodies were laid for their final rest. These chambers are buried beneath a massive, carefully constructed mound. The Neolithic view of death, however, was less absolute than ours today. The spirits of the ancestors were believed to have played an active part in the affairs of the living, and, indeed, there is good archaeological evidence that on important occasions families extracted bones, maybe even entire skeletons, from their small chambers within the tomb. They were then taken out and displayed, before being returned.[3]

The people buried within the West Kennet Long Barrow would have worshipped and attended ceremonies at Avebury. They would have lived in the Marlborough Downs, just outside the sacred site of Avebury itself, and would have built, maintained and understood it well. Being practising farmers they would have understood the subtleties of the land and how best it could be used. Their family farms and holdings may well have been hundreds of years old by the time the great stones were erected. These were people steeped in the landscape in which they lived and moved, a landscape that they moulded to fit their lives and the way they saw their position in the world. At Avebury you can still sense their fondness for what was, and remains, a very special place.

4

Great Orme Copper Mines

The Biggest Prehistoric Space in the World

I have long been fascinated by mines. In their own ways, they are just as impressive as castles or cathedrals: symbols of human aspiration, steps along the path of progress. I tend to look at them as a digger, as somebody who moves large quantities of earth for a living. And whenever I go down a mine I am impressed not just by the hard labour, but by the skill and also the courage and daring of miners, past and present. My digging life has been very different to theirs. For a start, I spent my years excavating out in the open air. Admittedly, the atmosphere in a gravel quarry could become dusty at times, and diesel fumes from bulldozers and excavators could get a bit much, but at least the air was constantly moving. Down in a mine, life would have been very different.

So far as ancient mines go, my favourite is to be found on the coast of north Wales, near Llandudno. The mine, or rather mines, lie deep within a large hill, the Great Orme, that overlooks the famous Victorian seaside resort (the largest in Wales) and the sea itself.[1] The arrival of the new technology of metal-working in Britain happened shortly before 2500 BC. We now know that it was also a complex social process that probably involved specialists from overseas. The first metal to be used for making tools was copper, but after about 2500 BC it was replaced by bronze, an alloy made up of about 90 per cent copper and 10 per cent tin. The new alloy was much harder and more durable than copper on its own, and soon proved very popular. Within a couple of centuries, it had led to the construction of some deep mines. Great Orme was one of them – and one of the largest in prehistoric Europe.

The hill at Great Orme is honeycombed by numerous shafts and galleries. They were first hewn out of the rock early in the Bronze Age, around 1900 BC and the mine then flourished until about 1500 BC, when production fell away, probably because there was so much scrap bronze in circulation. But mining at Great Orme didn't stop entirely and continued in a much-reduced capacity into the early Iron Age, eventually ceasing altogether around 600 BC. The hill was then left in peace for over two millennia, until, in the eighteenth and nineteenth centuries, copper mining resumed again in earnest. During this time, many of the Bronze Age galleries were cut through by new workings, and one wonders whether those Georgian and Victorian miners had any idea of the antiquity of the long-abandoned shafts that we know they encountered.

Today you enter the Bronze Age shafts at Great Orme from a modern quarry face, but in the past you would have clambered through a series of near-vertical and horizontal tunnels to reach the area being mined. Long journeys to the working face have always been a feature of life in a deep mine, in prehistory as today. Now, of course, the extracted ore from deep mines is taken to the surface by lifts and conveyor belts, but in prehistory it had to be carried out by those who had mined it. They had to take it all the way up to the surface, or along narrow galleries to a vertical shaft where it could be raised by a winch.

So far as we currently know, the Bronze Age mines at Great Orme can be traced for at least 230 feet below the surface. That is an extraordinary depth. There are many other reasons why the Bronze Age mines there are remarkable, but what has always moved me is the simple fact that when you enter a shaft, passage or gallery you are in a space that has been completely created by Bronze Age people. At Great Orme one becomes very aware that the achievements of the remote past can sometimes rival anything the present has to offer.

When you take a walk in a mine, you must look closely at your immediate surroundings. Apart from anything else, it's a good idea to do so if you want to stay upright: the ground beneath your feet is not always smooth or dry. On my first walk along a gallery at Great Orme, in 2002, I was accompanied by a resident archaeologist who showed me the thin veins of the distinctive green copper ore, malachite, which stood out clearly on the walls and ceiling. He explained how blocks of ore-bearing rock were removed by lighting fires and then throwing on water to quench the flames, causing the hot rock to crack and split off. I could also quite clearly spot the marks left by the stone hammers, and bone and antler tools that were used to break up the blocks of ore, and then to shape the gallery we were walking along. Here and there we could also spot smoke stains left by flames from the Bronze Age miners' lamps. I found the presence of those long-dead miners uncannily palpable. Despite myself, I had a strong urge to glance over my shoulder, in case anyone was there, watching us.

While you can walk along many galleries without bending, there are also tiny galleries that resemble rounded pipes or tubes more than tunnels, which could only have been worked by children. Today, child labour is nearly always the nastiest form of exploitative slavery, but did such conditions apply in the Bronze Age? There is evidence that a good number of the dozen or so Bronze Age mines known in Britain and Ireland, though probably not Great Orme in its heyday, were worked part-time. This would have happened in the quieter months of the farming year, from late autumn into winter. Children would also have been expected to work on the farm and in the home. Some would have been taught more specialized crafts and trades, such as metal-working and boat-building. Maybe those children who worked at Great Orme went on to become jewellers and metalsmiths. Of course, I may be entirely mistaken – but in most ancient societies, to put it crudely, children who survived the perils of early childhood were a valuable resource to the community, and their survival would have been essential to its future prosperity.

Many of the underground galleries at Great Orme open into at least one of two huge underground chambers. By torchlight it's hard to get a true impression of the size of these vast caverns, but during filming a few years ago we floodlit one of them: it was, I guessed, at least sixty or seventy feet high and probably twice as wide. Its vast, irregular walls were honeycombed with openings into, and along, additional galleries. The space itself was so big and so irregular that one's voice was absorbed. I expected an echo, but there wasn't one. I have been told that the largest of the two chambers is the biggest enclosed prehistoric space, anywhere.

There is one big difference between modern and ancient mines.

Essentially it boils down to economics. In prehistoric times, work was not driven by profit: the market economy had yet to evolve. Instead, people did things to comply with family rights and obligations. A young man might find he owed a quantity of ore as 'bridewealth', a reciprocal exchange with the family of his new young wife. So while working a mine may have been dangerous and difficult, it was not done for the exploitative reasons that sometimes applied in the industrial era. Perhaps it might be an exaggeration to say that everybody working inside the hill at Great Orme was there by choice, but it strikes me that in prehistory the fundamental motives behind such Herculean tasks – including the shaping of the great stones at Stonehenge – were ultimately to do with what the miners' families expected of them. They were tasks that proclaimed the success and authority of a particular family, and the younger, fitter and stronger people were happy to perform them.

Offerings are frequently found in prehistoric mines. Sometimes these can be quite elaborate; more often, they are simple.[2] At Great Orme they most commonly took the form of small heaps of meat bones. These were once the remains of meals that were left on the floor, intended perhaps for the spirits of dead miners or the deities that inhabited the mine. Such offerings provide clues about the miners' ultimate motives and intentions. They suggest that the work was undertaken voluntarily, rather than as part of a slave-labour scheme – in which case one might have expected to find remains of weapons, chains, whips and the other accessories of slavery.

The men in the mines were also clearly aware of the natural dangers surrounding them. At Great Orme, flooding was a constant menace. So the miners constructed drainage channels, or conduits, to siphon water out of the galleries. Standing in that immense chamber at Great Orme for the first time, I felt overwhelmed by the sense of the lives that were lived here over the millennia. I became aware of the faint sound of water dripping, somewhere far away in the background. Then slowly it dawned on me: that very sound would have been heard by Bronze Age miners, over 3,000 years ago. I was standing in their space, listening to their sounds. Who needs time-travel?

5

A White Line in Time

Hadrian's Wall

The facts of Hadrian's Wall speak for themselves. Construction started in AD 122, following a personal visit to Britain by the Roman Emperor Hadrian. Today, the Wall runs for seventy-three miles across the uplands of northern England, linking England's east and west coasts; in places it survives some thirteen feet high. But it is not just a wall. It is a complete defensive system: think of the Maginot Line of the 1930s, or the hastily constructed defensive lines erected in Britain in late 1940 and early 1941. The Wall featured major strong points: sixteen massive forts, plus numerous smaller camps and mile-castles. There is also a network of supply routes and additional outworks, such as banks and ditches. More to the point, the Roman wall and its modern equivalents were not just about defence. They were also meant to proclaim comfort and security to the communities they were intended to protect. And they symbolized defiance and power to all who would threaten them. Stretches of Hadrian's Wall were painted white to make them more visible and impressive.[1]

I have to confess that in the past I was never very fond of Britain's brief Roman period, from AD 43 to about AD 410. This was largely, I think, down to the way it was taught. Stressing the militaristic and political, it was all about the doings of the Roman Army and its manifold units with their arcane names – *legio VI Victrix* and so on – and the various forgettable emperors and generals who commanded them. Acres of print were devoted to their countless camps and forts, together with those lavish villas and their under-floor heating. We seemed to learn very little about the lives of ordinary Romano-Brits, although we did discover that towns in Roman Britain never really caught on. But why was this? Were Romans and

Britons friendly? How did they relate to each other? Was society divided? What happened to the Britons' Celtic identity? These were the questions my student self wanted answered.

On reflection, perhaps I and my contemporaries were reacting to a childhood spent in the shadow of the aftermath of war. In my memory, the 1950s were drab and colourless: food rationing was still in place; uncles and cousins went off to do National Service; military convoys were ubiquitous. Service men and women were everywhere and then, slightly later, there was the insanity of the Vietnam war – which, like all my friends, I bitterly opposed. So I think it entirely unsurprising that I lost all interest in a view of Roman Britain that seemed to mirror everything I disliked about contemporary Britain.

Then in 1989 I came across a book that changed my attitude to the Romans and to Hadrian's Wall.[2] Almost immediately I was gripped. The author, Stephen Johnson, whom I knew very vaguely from academic conferences, had treated the Wall as something built by human beings in a real world of varied landscapes – fields, scrub, trees, rocks, rivers, fells and hills.[3] I had to see it for myself.

So I lost no time, and headed north. Archaeology has always been a small world, and I was able to visit various excavations run by friends and colleagues. Soon I began to get a real feeling for the place. I could understand why the Emperor Hadrian had felt it necessary to build his Wall. It was all about consolidating a frontier zone. But it wasn't a simple division of Civilization on one side and Barbarism on the other. Far from it. We now understand that some of the landscapes through which the Wall passes had been farmed and partitioned into fields during the Iron Age, and even earlier. So it was more an assertion of the new Roman identity than a simple military defence. It was proclaiming the arrival of *Romanitas* (or Romanness) – what today we would regard as classical civilization – as opposed to the less formal ideals of Celtic culture.

Hadrian, who was emperor from AD 117 to 138, started work on his Wall in 122 and continued to build it, together with its various forts and mile-castles, for the following twenty years. It was probably never completed, in the sense that we would understand the term. Excavation at various places along the Wall has revealed that it had a complex history, and for almost ten years in the mid-second century it was abandoned in favour of the shorter, turf-built Antonine Wall that ran across the central belt of Scotland, between the Firths of Clyde and Forth.[4]

Over the past half-century research has been focused on some of the great forts at Corbridge, Vindolanda, Birdoswald, South Shields and Wallsend. The results of these digs have given us fascinating insights into the real lives of soldiers stationed on the Wall. In particular, wooden writing tablets found at Vindolanda, and now on display at the British Museum, have in effect given us examples of 'letters home', from soldiers to their friends and families, often abroad, and their replies.[5] One tablet famously includes an invitation to a soldier to attend a birthday party in about AD 100 (this is also probably the oldest known Latin writing by a woman).

The Wall would have passed through largely empty landscapes in some of its higher and more rocky regions, but it was by no means all like that. For much of its length, the people manning the forts and mile-castles would have looked out over fields and settlements. Indeed, the construction and maintenance of the Wall would have brought considerable prosperity to the region and recent research suggests that this was to have a long-lasting effect. The area continued to thrive after the withdrawal of the Roman Army in the early fifth century. It had acquired a life, and culture, of its own.

The reconstructions (for example at South Shields) are remarkable, but almost by definition, they leave less for the imagination, which is why I generally favour some of the lonelier, upland reaches. Again, it's a personal choice entirely, but the forts I have really enjoyed visiting, and have returned to from time to time, are those at Birdoswald and Housesteads. The area I know least well is the defended stretch of the Cumbrian coast south of Bowness, where the main Wall ends. I have flown over it, while filming *Britain AD*, and was able to spot a number of forts, fortlets and towers that overlooked the sea.[6]

I spent a couple of years in the area south of the Solway Firth, as a child of twelve and thirteen. Sadly, it was not a very pleasant or happy experience, which might explain why I have not rushed to explore the region more thoroughly. I have memories of distant views across the sea. On a clear day I could just see Scotland on the far horizon, as I took my long solitary walks to find peace in an often turbulent place. The wintry landscape, with dwarf, wind-warped sessile oaks fringing the hills and deep valleys, was strangely comforting – emotionally warmer than some of the people I was then with. I also gained strength there. It was where I first began to observe and appreciate my surroundings, and although I was far from content, those walks helped me put my problems into perspective. I suppose one should not allow ghosts from the past to spoil one's enjoyment and appreciation of the present, but in some respects those personal, lingering presences, which characterize certain places, also heighten one's reactions. They add power and contrast to perception: often with brightness and light, but sometimes with darkness, too.

6

Shifting Sands of Time

Seahenge, Brancaster and the Southern Wash

The beaches of north Norfolk and the Wash, with their seals and peat beds, samphire and mussels, are constantly changing. In some places, especially in the marshy landscapes west of Dersingham, when the tide's out you can barely see the sea, which can be almost a mile away. Sometimes the tide returns gently and you can stay ahead of it at a steady walk; at other times – and this can be quite unpredictable and not always dependent on the weather – rip tides rush in at a headlong rate.

I first became acquainted with this region when I became involved with the excavation of a highly unusual site at Holme-next-the-Sea, a small seaside village about three miles north-east of Hunstanton along the coast. Holme lies on the south-eastern side of the mouth of the Wash, about twelve miles south-east of its Lincolnshire equivalent, across the Wash at Gibraltar Point, a name familiar to aficionados of Radio 4's early morning shipping forecast. Today its coastline is dominated by healthy stands of Corsican pines, planted by the nearby Holkham Estate in the early twentieth century, and features a wonderfully atmospheric nature reserve (the Holme Bird Observatory) with a huge population of unusual waders – as well as a large visiting population of bird enthusiasts, or birders, instantly recognizable by the binoculars slung around their necks, the expensive telephoto lenses on their cameras and a complete inability to talk about anything other than birds.

In the early spring of 1998, John Lorimer, a local man, made a most remarkable discovery on the beach at Holme, just above the low-tide line. It took several tides and repeated visits to become completely clear, but eventually it revealed itself to be a small circle of fifty-five posts surrounding an inverted oak tree. Many of the posts had been carefully split in half and there seemed little doubt that this structure was a miniature timber circle and belonged to the same Bronze Age tradition as round barrows and henges. Thanks to the technique of dendrochronology, or tree-ring dating, we soon discovered that the central tree, together with several of the posts which surrounded it, had been felled at precisely the same time: between April and June in the year 2049 BC. Presumably the little shrine, or monument, had been constructed very shortly thereafter.

The constant danger posed by the sea was very evident during the excavation of what became known as Seahenge. A second, more complex circle, or circles, of posts and smaller woodwork, such as stakes and poles rather than posts and planks, but constructed at exactly the same time, was found close by.[1] These rings of woodwork surrounded two short, shaped wooden bearers for a massive wooden coffin, which has long since been claimed by the sea. The first timber circle, we believe, served as a shrine, where the friends and family of the dead person held a funeral. Once the ceremony had been completed, the body was carried across to the second circle, where it was placed in a coffin that rested on the two wooden bearers. The coffin was protected for many millennia by a mound (or barrow) of soil and mud that eventually succumbed to the encroaching waters of the North Sea.

I find the fragility and constant changing of the north Norfolk coast exhilarating. Things happen so fast here. Initially Seahenge would have been protected behind a low barrier of dunes, which probably eroded away in post-Roman times as the level of the North Sea continued to rise. After that, the timber circle would come and go as high tides and passing storms would shift the sand. The Iron Age peat beds that surrounded it on the beach, were constantly eroding, too. A few miles to the west, the Scolt Head dunes – now a National Nature Reserve – probably appeared after one or two major storms, if their Dutch counterparts are anything to go by. Prehistoric communities would have adapted to these changes by moving somewhere less at risk. But how did the later Romans cope with such a changeable landscape? They had investments in the area: an army and civilian settlements to protect. How did they fare?

If you take the coastal road east out of Holme-next-the-Sea, you come to the small village of Brancaster. Anyone with an eye for place-names could spot the name was derived from Latin (the 'caster' bit is the giveaway, from the Latin *castrum*, meaning a military camp or station; 'Bran' refers to its earlier Celtic name).[2] Brancaster was the most northerly of the eleven so-called 'Saxon Shore forts', Roman defences that extend around the south-east coastal approaches of East Anglia, the Thames estuary, Kent and Sussex to end at Porchester, near the Isle of Wight.[3] The conventional wisdom used to be that these forts were built to repel the continental invaders who would later be known collectively as the Anglo-Saxons. It's a straightfor-ward enough explanation, but unfortunately history is rarely that simple. The latest evidence suggests that there were no massive Anglo-Saxon invasions or consequent wholesale population displacement in Britain. Instead, the changes that we see in the infrastructure of later Roman Britain were brought about by increasing prosperity through-out the fourth century AD, which in turn led to greater contact with continental Europe and with it, of course, a degree of population movement: maybe a quarter or a third of what would soon be the English population changed over a couple of centuries. But it was a two-way process: yes, people from Germany settled in Britain, but Britons crossed the North Sea and settled on the Continent.[4] The idea of large-scale invasions is difficult to support, quite simply because there is so little evidence for conflict in southern Britain during the early post-Roman centuries. Yes, there were changes, particularly in the rural economy, where cereal-growing was replaced by mixed farming, with a far greater reliance on livestock. This was still a pros-perous economy, however; settlements were growing in size, and field systems continued to expand.

So what about Brancaster and those forts of the 'Saxon Shore'? Were they part of a defensive line, like Hadrian's Wall? It was an explanation I had long been unhappy with, not least because the Saxon Shore forts were strangely diverse, both in their layout and in their positioning along the coast. Maybe it was this costal position which favoured their survival, but the Saxon Shore forts are some of the best-preserved Roman structures in Britain, with stout masonry walls that still tower above the visitor. I'm particularly fond of the forts at Porchester Castle, on the Solent in Hampshire, and Burgh Castle, further south along the Norfolk coast from Brancaster. On excavation, some of the southern forts appeared more like well-defended warehouses than garrisons. This was in huge contrast with what we knew and suspected about the Norfolk forts.[5]

Then, early in 2012, my involvement with this north Norfolk landscape took an unusual twist. For many years I had been filming with the long-running Channel 4 series *Time Team*. Series 20 was to be the last one, with eleven new episodes. The last one of all would be filmed at Brancaster. Brancaster was a Scheduled (i.e. protected) Ancient Monument and a few weeks earlier I had signed the official consent forms that would allow us to dig there.

I had first visited Brancaster in 1977, to take part in excavating a civilian settlement, just to the west of the fort itself. So I knew a little about the site before we started the *Time Team* dig. When we began in 2012, we assumed that the settlement dug in 1977 was accommodation for the soldiers' wives and families, who were not officially allowed to live inside the fort. What we then found surprised us all. The many coins and potsherds that we unearthed within the fort dated to the second and earlier third centuries AD, long before the supposed Saxon threat of the fourth and fifth centuries.[6] Aerial photos also revealed another huge civilian settlement on the eastern side of the fort. Whatever else may have been going on, Brancaster was not looking like a hasty late Roman defensive measure.

The atmosphere on that final *Time Team* dig was strange. It was such an honour and a thrill to be allowed to excavate within the walls of the Scheduled area of the fort, but we had just three days to make sense of a complex site: a daunting proposition. Soon, however, the ground began to reveal its secrets and our sense of urgency returned. The Roman fort of *Branodunum*, we found, was much more than a last-minute defensive structure built to exclude Saxon invaders. It turned out to be a major military installation whose roots lay early in the Roman period. Providing protection for a small harbour, it was home to a cavalry regiment whose enormous, roofed, oval training arena, or *gyrus*, together with the huge buttresses needed to support such a large roof span, was revealed by the ground-penetrating radar of *Time Team*'s geophysicists. I'll never forget seeing that distinct shape on the screen.[7] It was so clear. I'm happy to say that it's still down there, deep underground, buried beneath six feet of sand, earth and rubble, an unweathered and intact major Roman building. I know it isn't happening, but I cannot walk across that site without seeing a vision of deeply buried Roman bones, pottery and glass crunching beneath my feet.

The final twist in this story took place when I recently drove around the area to refresh my memory for this book. It was November 2016, and I was heading west along the coastal road, towards fish-and-chips at the nearby port of Wells, when I passed the orchard where in the autumn of 1998 we had built the full-size reconstruction of Seahenge for the first *Time Team* documentary about it. It had been a major effort, as we had used Bronze Age replica tools throughout, and at the end of the final day of filming, the producers had organized a party. I remember it well: as the laughter and general hilarity grew, I found myself walking away from the crowd, towards the small ring of oak posts we had just finished erecting. The sun had set, but there was still a glimmer of light, low on the horizon. A breeze had sprung up, as often happens along the coast in the evening. I slipped sideways through the narrow entranceway into the close circle of split oak posts.

Seeing it again, and on my own, the rebuilt Seahenge was not at all what I had remembered. It was subtly, but significantly different. Now that the many cameramen, sound recordists, archaeologists and helpers had left, the silence was, to use a cliché, deafening. When we first built it there had been an incredibly strong – almost literally overpowering – scent of tannic acid from the freshly split timbers. The smell was so pungent that I could feel my skin start to tighten and my eyes to run. Inside the timber circle, the fading daylight was largely excluded by the tall walls, as were sea breezes and all noises from the nearby party. As I cleared my throat, the timbers reflected and distorted the sound, as if I were in an empty church or hall.[8] I became aware of being very alone and a little frightened. For a few moments I genuinely felt I was on the edge of other worlds. Then slowly I regained full self-control. It was the strangest of experiences.

As I slowly walked back towards the party, I suddenly changed my mind. After that experience, I somehow couldn't face a crowd of jolly people. So wearily I climbed into the Land Rover and headed home, tired but profoundly grateful: that vivid glimpse into the Bronze Age would take time to digest. It's a process that is still a part of me and I hope it never fades.

7

Arthurian Tintagel

Myths and Realities

The precise moment when the Roman occupation of Britain ended remains shrouded in mystery. We do know, however, that the Roman Army was being withdrawn from the Roman province of Britannia in the early years of the fifth century AD, and most historians agree that the province itself ceased to exist sometime around 410. But we also know that, only the century before, Britain south of Hadrian's Wall had been flourishing, acquiring its own distinctively Romano-British identity. Two centuries later, that identity had become completely transformed: by AD 700 people spoke a different language and some were starting to think of themselves not as British, but English. It must have been a time of extraordinary rebirth and renewal, and not remotely a 'Dark Age' (as the post-Roman centuries were conventionally termed). To describe what was happening in the south-west of the country between 450 and 650 we now use the term 'Early Christian'; in the midlands, south and east, the preferred name is 'Early Saxon'. Archaeologists are getting very much better at identifying new sites of these periods, but some places have always had a reputation. Of these, by far the most famous is Tintagel, the supposed birthplace of the legendary King Arthur.

Tintagel is a small rocky promontory on the north coast of Cornwall, with remains of a thirteenth-century castle and a sheltered sandy beach, known as the Haven. The surviving walls of the castle have to be among the most romantic ruins in Britain. It's not the – rather ordinary – medieval stonework that conveys the atmosphere, but the castle's setting: cliffs, jagged rocks and the constant angry presence of the sea. This landscape, like others in the south-west, has never been smoothed over and rounded by glaciers, so the rocks still appear sharp and jagged. It is a place that has been the subject of myth and Arthurian mayhem ever since *c.* 1136, when Geoffrey of Monmouth wrote his *Historia Regum Britanniae* (History of the Kings of Britain).

I don't think Geoffrey chose Tintagel at random. It is such a magical and evocative place, and not just in the wild and stormy months of autumn and winter. I can remember dropping off to sleep during a break when we were filming *Britain AD* for Channel 4, during the summer of 2003. I was lying in short grass by the edge of the steep drop down to the sea. I could hear the gentle sounds of summer waves breaking on the small sandy beach in the Haven, far below, while all around me the sound and sight of wheeling seagulls slowly lulled my senses. Soon I was fast asleep, and then, of course, I dreamed. Of Arthur. Had I been living in the Middle Ages, I would probably have told people I had been visited by the spirit of the man himself. I know I felt strangely disembodied when the assistant cameraman woke me up. It is indeed a very strange place.

The promontory or headland at Tintagel is almost an island, but not quite. So its position is well defended against surprise attack. The first attempt to build on these natural defences was the Great Ditch, which probably dates to Early Christian times and was placed to cut off access to the headland. The medieval castle of Tintagel consists of the 'island' itself, together with fortifications near the Haven, plus walls on either side of the bridged crossing to the mainland. These walls were built after 1233, following his acquisition of the land by Earl Richard of Cornwall. Why did he choose this site? Yes, it is an easily defended promontory, but it is hardly in a strategic position of any military importance. It isn't like Windsor Castle, defending a crossing of the Thames. The answer to the question very possibly lies in the Great Ditch and the two walls that, in the thirteenth century, would have immediately struck the visitor. It was also a known stopping-off point for coastal traders who were seeking up-market, prestigious customers. More probably, however, it would have been the place's Arthurian reputation and maybe, too, memories of even earlier glories – which I will return to shortly. I am in little doubt that it would have been the sort of place that any powerful man would have wanted to control.

So now to the mythical king.[1] The first full account of the deeds of King Arthur is that by Geoffrey of Monmouth. He probably based some of his stories on another mythical Welsh king, also called Arthur, who appears in an anonymous account, *The Early History of the Britons*, written and put together around 830. This early collection, very anti-English (or rather, anti-Saxon) in tone, was seeking to establish the legitimacy of Wales as an ancient British nation. But Geoffrey's later book proved more popular and enduring. It captured the essence of courtly chivalry, and Arthurian legends were to be a feature of the Middle Ages, both in Britain and across the Channel. The culmination of these retellings of King Arthur's deeds came some three and a half centuries later, with Sir Thomas Malory's *Le Morte d'Arthur*: written in the late 1460s, it was printed and published by William Caxton in 1485. Malory's book was the inspiration for many more retellings, right up to T. H. White's 1958, *The Once and Future King*.

Arthur's conception and birth at Tintagel, as portrayed by Geoffrey of Monmouth, was very strange. His father was the legendary Welsh king, Uther Pendragon, who was disguised by Merlin's magic to resemble his worst enemy. In this disguise he slept with Ygrain, Duchess of Tintagel, his enemy's wife, at her castle. The result of their union was Arthur. In the Middle Ages, myths and legends were reinforced by physical objects, such as pieces of the True Cross, or the bones, or limbs, of saints. But in 1191, King Arthur's corpse was revealed, thanks to another king – this time a real one – Henry II (1154–89): the body was 'discovered' by monks at Glastonbury Abbey, just two years after Henry's death, and most probably on his suggestion. The remains of Arthur, who was 'found' with his wife Guinevere, were immediately buried in an elaborate tomb at the abbey, following which, thousands of visitors flocked to pay their respects[2] – unsurprisingly, this had a galvanizing effect on the abbey's finances. It also transferred the magical power of Arthur from Wales to England, which was probably Henry's intention from the outset. As political PR moves go, this has to have been a master-stroke. It's just the sort of romantic fantasy that social media today would have seized upon.

Of course it's easy to dismiss such legends as 'false news', but we should also try to see them for what they were: stories that gave the ruling powers their legitimacy, popularity and authority. Again, nothing changes. But they did also help support monasteries and the Church, not to mention the many inn- and shopkeepers who served pilgrims in places like Glastonbury.

I mentioned earlier that the real facts behind Tintagel are just as remarkable as the legends, and there are some fairly obvious clues to help us reach them, including the Great Ditch, the standing castle walls and some mounds and very low ruins on the promontory itself. These would have fired the curiosity of any visitor who knew about archaeology. And one of those was the distinguished Professor Ralegh Radford, who undertook a series of excavations from 1933 to 1939. He revealed the walls of rectangular buildings (which are still quite clearly visible as low, grass-covered banks), which could be dated by pottery to between the fifth and seventh centuries AD. The pottery itself was remarkable as it included many fine pieces from the eastern Mediterranean. These would have been landed by ships berthing at the Haven and we now know that Tintagel was part of a regular trading network that linked western Britain with Iberia and the Mediterranean. The main export was tin – as in Roman times.

So Tintagel was clearly a very up-market, elite settlement, where people were feasting and drinking imported wine. It has even been suggested that it may have been a royal centre for the fourth- to eighth-century Kingdom of Dumnonia, which included Devon, Cornwall and parts of Somerset. This seems to me entirely reasonable. Legends rarely appear out of the blue: there is usually something to inspire them. I was also delighted to discover that in 2016 English Heritage started the first season of a new Tintagel research excavation, which readers will be able to follow on their blog.[3] Already the results look very promising, with more exotic imported pottery and good evidence for large buildings.

I find the links between the medieval manipulation of Arthurian myths and the political realities of the time deeply intriguing. The stories echo and enhance the prevailing social and intellectual taste for chivalry. So were they just cold-blooded manipulations of established myths, to achieve what were, ultimately, political ends?

Perhaps. But I find the Arthurian story fascinating for the insights it gives us into the surprising sophistication of medieval power politics and public relations. These people were spin doctors. They knew only too well how to 'manage' and manipulate their supporters. Those jagged, romantic rocks of Tintagel are ultimately telling us a complex and politically sophisticated story.

8

A Haunting Place

Whitby Abbey

My first encounter with Whitby, on a late autumn morning in the mid-1980s, was almost accidental. I was on my way home, after staying at my sister's sheep farm, high on the Yorkshire Moors, when I was practically stopped in my tracks by the gaunt ruins of the Abbey Church that commands the eastern approaches to the town. I had been driving for some time through the open Moors with their drystone walls and high common moorland pastures. It's an ageless landscape where, you feel, an encounter with the inhabitants of *Wuthering Heights* would come as no surprise. Although much lower-lying, the coastal town of Whitby somehow manages to remain a part of the Moors. It retains the same open and rather wild atmosphere, which hasn't even been slightly tamed by the Abbey. The town and port enclose a natural steep-sided harbour at the mouth of the River Esk. The streets of the old town follow the contours of the land and are narrow and winding. There are numerous footpaths and flights of steps. The Abbey and the parish church are on more open ground on the east side of the river. The church overlooks Whitby, I always think in a friendly, almost paternal fashion. The Abbey, by contrast, seems to dominate it.

Whitby has long been a small commercial harbour and a successful fishing port. It is also well known for its superb Abbey and nowadays too for its oak-smoked kippers, to which I am addicted. I honestly don't know if it was the Abbey ruins, or the kippers, that first got me hooked on the town and its surrounding landscape. But Heathcliff and Cathy were not the only literary figures that might have been seen in nineteenth-century Whitby. The town's most famous, or rather infamous, fictional inhabitant was far more scary.

Bram Stoker's hematophagous Count Dracula landed at Whitby, after a voyage from land-locked Transylvania in the *Demeter*, a Russian vessel that was carrying a cargo of silver sand and boxes of earth. It had been an eventful voyage, one which had involved the slow death of the entire crew, consumed by Dracula, whose undead body lay in a coffin below-decks. Eventually all were dead, except for the captain, who was lashed to the wheel. On 8 August the ship ran aground. Taking the form of a large dog, Dracula leapt from the ship on to the beach below East Cliff, near the ruins of the Abbey. He was now free to indulge in his horrifying blood-sucking adventures, which have scared Stoker's readers stiff ever since.

Although Whitby is hugely popular, especially in summer, tourism doesn't seem to have destroyed its character: it has the feel of a thriving community. The harbour is still home to active fishing boats and even the big tall-masted sailing ships, which are moored here for the benefit of visitors, are all sea-going vessels. But Whitby is a small town and the noise and bustle of the harbour is quickly left behind. On one occasion, when we were filming *Britain AD*, we took our equipment to the edge of town to record some quiet ambient sound. We found ourselves beside the outer approaches to the harbour, on a stretch of sandy beach along the eastern bank of the River Esk. It was early evening: shouts of children playing far in the distance, noisy seagulls overhead, waves washing the shore.

On our way to the location, the sound-recordist had been worried. For various reasons he hadn't been able to capture much by way of background sound – what soundmen call 'wild track' – but this beach seemed ideal, and he started to relax. We sat on a low timber groyne; I held up the microphone while he focused on the dials before him. Seconds passed. Suddenly, there was a deafening bang. Instinctively, I ducked. Then another one, even louder. This was followed by a short pause, before two answering explosions came from the opposite bank of the river. Then the cause of this mayhem revealed herself: a magnificent three-masted tall-ship, gently easing her way out to the North Sea. Her cannons had been bidding farewell to Whitby.

This ceremonial salute put me in mind of another, rather more destructive barrage. In 1914, two German battlecruisers shelled the

long-ruined Abbey Church, high up on its promontory above Whitby.
The ships were thought to be aiming for a nearby Coastguard signal
station, but the shells from their large-calibre guns did considerable
damage to the Abbey ruins.

Whitby Abbey was founded in AD 657 by King Oswy of Northumbria, who appointed a remarkable woman, Lady, later Saint,
Hilda, as its first abbess. Of noble birth, Hilda was the grand-niece
of Northumbria's first Christian king, Edwin. In 664 Oswy convened the still-famous Synod of Whitby, where it was agreed that the
Church in Northumbria – in effect all of northern England – would
leave the Celtic Church tradition and join that of the Roman Church.
The Synod is therefore seen as an important step in the Romaniz-
ation of the Church in England, and a major threshold in the growing
unification of England itself, as it helped to increase central, and
with it royal, authority. Then disaster struck.

Now I have always been very interested when I hear, or come
across, anything about the Vikings in England. This is partly due to
the fact that I look like one, and I have always been aware of a farm
near our family home in Hertfordshire, named Dane End. So about
ten years ago, I had my mitochondrial DNA tested and learned that,
like many people from the east of England, I was descended from
Vikings. I have to admit this came as no surprise, as my blood group
is A (+), which as a student I was taught was linked to Viking incursions into eastern England and northern Scotland. And Whitby
didn't escape these incursions. Devastated in a series of Viking raids
between 867 and 870, it lay in ruins for the next two centuries. In
1078 the Abbey was re-established as a Benedictine monastery,
before being dissolved in 1539. It's the towering, yet brooding (or is
that Dracula affecting my judgement?) ruins of this Norman structure that we can still see today.

I have always been intrigued by the secular buildings that replaced
abbeys after the Dissolution, and I think it a shame that they are not
given more prominence. Some can be magnificent, such as Fountains
Hall, the superb mansion built after the dissolution of Fountains
Abbey, some sixty miles inland from Whitby. It was most probably
designed by the Elizabethan/Jacobean architect John Smythson and
is a building which features huge rectangular windows which seem

to have been deliberately fashioned to contrast with their arched and arcaded monastic neighbours'. It's a four-storeyed building that proclaims its secularity, but with just a hint of the military, in the castellated towers on either side of the grand front. Although the man who commissioned Fountains Hall wasn't particularly pleasant (he was a Catholic-detector/chaser), his house seems aspirational, and yes, modern.[1]

Abbey House at Whitby is less spectacular, but it has a colourful history. It was built for the Cholmley family, who bought the Abbey after its dissolution. They were Yorkshire landowners with plenty of money, which was just as well, because the house fell into the hands of the Parliamentarians and was looted, following the surrender of Whitby in 1645, during the English Civil War. In the 1670s a two-storey and severely classical ten-bay Banqueting House was added. Today this houses the Abbey visitor centre. The Cholmleys moved out in the early eighteenth century, whereupon a powerful storm severely damaged the house, which lay abandoned for most of the eighteenth and nineteenth centuries. This was the time when the Abbey ruins, completely exposed to ferocious gales from off the North Sea, began to crumble quite rapidly. The central tower of the church collapsed around 1790.

One of the things I find admirable about mid-seventeenth-century England is the way that its towns, villages and people dealt with, and recovered from, the animosity and conflict brought about during the Civil War, whose aftermath saw the establishment of a constitutional monarchy and a less top-heavy, aristocratic social system in which a new middle class came into being. The economic and social transformation was achieved through increased trade, industry and commerce. And Whitby played its part. The port expanded throughout the eighteenth century and shipbuilding, using locally grown oak, became a major industry. By the end of the century Whitby was the third largest shipbuilder in England (after London and Newcastle). It was also an important centre of the highly profitable whaling industry.

Close by the ruins of the great monastery, and slightly nearer to the sea, is the medieval Church of St Mary, which was used by the parishioners of the town. I had seen it many times on my trips to the nearby Abbey, but it took me a while to get round to putting my head through the door.

A fine medieval building, St Mary's is best approached by way of the 199 Church Steps that head up the hill from Church Street in the old part of town, below. It has an extraordinarily complete Georgian interior, from the time of the town's greatest prosperity. Some medieval windows have been replaced and given an unexpectedly domestic Georgian twist. The elaborate, elegant Georgian galleries and box pews are still intact. One could imagine a service here in the eighteenth century: the packed congregation and the rustling of silks. I know no church interior like it and I suspect it is my favourite, not because of its architecture so much as its ability to transport any visitor back to another age. It is exceptionally atmospheric. The memorials to lost crews, fishermen and seamen that grace its walls still communicate directly their families' grief. The wooden furniture and fittings look real and somehow 'lived-in'. It's as if a Georgian congregation had just left after morning service and were heading back down the Church Steps to their Sunday lunches, in the town below.

A few months ago, I used the writing of this book as an excuse to revisit St Mary's. After I had spent an hour or so in the church I strolled back to the car, opened the boot and took out the relevant Pevsner (*Yorkshire: The North Riding*), which was one of the original editions, from 1966. Normally the great man is very restrained when it comes to parish churches, but not here. He describes it as 'impossible not to love'.[2] How right he was.

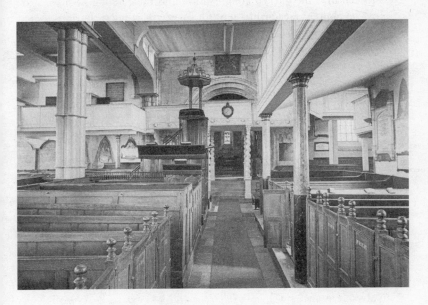

Whitby parish church.

9

The Scottish Borders

The View from the North

The sense of a landscape also depends on where you're seeing it from. To be more precise, I have always viewed the Scottish Borders from an English perspective, and this despite the fact that half my family has roots north of the border (I would tell you my mother's maiden name if it didn't undermine my digital security). In my experience, roots and ancestry take second place to education and upbringing – and I suspect it was ever thus, for many people.

To me Hadrian's Wall is not just an archaeological monument, it marks the southern edge of the Borderlands. Although my rational self has long told me I was wrong, my emotional self continued to view the country north of Hadrian's Wall as hostile outlaw-land. Then one day, about twenty years ago, I decided to take a closer look over the edge of my own little conceptual box labelled England. And what I discovered entranced me.

Of course I had visited Scotland before that moment, many times. But for these visits, mostly to Edinburgh, Glasgow and Orkney, I had travelled either by air or by rail, and although a long tunnel wasn't involved it might as well have been for all I was aware. As I recall, it was a day in September, 1995 or '96, and we were on a visit to archaeological sites in the north, possibly when researching for an idea that would later evolve into *Britain BC*. We were driving north along the Great North Road (the A1) and decided, on a whim, to fork left just outside Morpeth onto the A697. I wanted to follow the A1 along the Scottish east coast, but Maisie was definite: she wanted to head due north. As roads go, this was certainly 'one less travelled'. It went relentlessly uphill and the houses, trees, hedges and banks around us slowly dropped away, to be replaced by drystone walls, distant small farms and sheep. Maisie was delighted, as it was the route her Scottish father used to follow when they stayed with the family in Moray, northern Scotland, on their summer holidays. Free from the traffic of the A1, we were now travelling through the sort of open uplands that are familiar to anyone acquainted with Hadrian's Wall. You could call it 'classic' Northumbrian hill country. As a sheep farmer, I think of it as the home of the hardy, white-faced Cheviot breed.

We continued to head north, and as we started down the slope into the Tweed valley, I was amazed at the luxuriance of the landscape, the lush greenness of everything: such a vivid contrast with the country we had been travelling through, to the south. The valley seemed more populated too: there were more farms and houses. We crossed the Tweed, which forms the border between England and Scotland, at Coldstream, a town that has given its name to perhaps the most famous regiment in the British Army.

The valley of the River Tweed has a character all of its own, whether in England or Scotland. There are villages, small market towns and many farms. The countryside features fields and meadows, grazing and arable, together with woods, spinneys, hedges and copses. It's not typical of the bleak, open and much-fought-over uplands of the Borderlands, as portrayed by romantic novelists and many Hollywood film-makers. Landscapes in reality are often more complex than their popular images. But surely, and this is something I've heard repeatedly throughout my professional life, weren't things simpler in the past? I don't think so: quite the reverse, in fact.

I knew that the Scottish Borders had a very rich prehistoric past, and I was also aware that the area enjoyed considerable prosperity in Roman times. More recently it has been shown that this prosperity extended into the post-Roman centuries, too. Analysis of ancient pollen and plant remains has shown that the landscapes around Hadrian's Wall did not revert to impenetrable woodland at the end of the Roman period. Fields and grazing continued to be used and maintained. Celtic Christianity was introduced to Scotland by St Ninian, who founded a monastery at Whithorn, near Newton Stewart, in Dumfries and Galloway in the early fifth century. Excavations there have revealed imported Mediterranean pottery of the fifth and sixth centuries, closely similar to finds from Tintagel and other trading posts along the western approaches.[1] These finds, and other evidence, suggest that Whithorn was far from isolated and was part of wider trading and cultural networks.[2] Fine abbeys are very much a feature of the Scottish Borders at Coldingham, Dryburgh, Jedburgh, Kelso and Melrose, so we should not characterize the entire area as war-torn and somehow dysfunctional simply because it faced disruption for some three hundred years, from the fourteenth to the sixteenth centuries.[3] Raiding and rustling were never welcomed, but the fact is that people coped with and adapted to these conditions. Only the largest-scale, bloodiest attacks had the power to bring daily life to a halt and these, mercifully, were relatively few and far between.

We tend to think of the Middle Ages as being somehow more lawless and less regulated than our own times, but I think we should be careful in jumping to such conclusions. If anything, their rules and conventions could often have been stricter: being framed within the context of the tribe and family, transgressors could expect rapid, often summary justice. National borders mattered far less than a family's or clan's sphere of influence, which could change from one generation to another. So to what extent is our understanding of the medieval Scottish Borders, as a blood-soaked landscape of free-wheeling outlaws, justified? I think the answer lies in the landscape itself, and how it was farmed. Cattle, and later sheep, formed the basis of the regional economy; but as forms of wealth go, cows, bulls and bullocks are very mobile – and easy to steal. So first a word about rustling.

Cattle rustling normally happened at a convenient time to the rustlers, in other words, at quiet times of the farming year; and it was always about more than just stealing. Very often it was carried out by young men, who did it to prove or show off their manliness – either to impress suitors or as part of a play for political power. Most livestock farmers would have taken special measures to look after the animals they valued most highly. Very often these would be prize bulls or rams or the best breeding cows and ewes. These would be the beasts that were given extra protection, sometimes in the lowest storey of a tower house. Today, these tower houses remain the defining feature of the Border landscape, a reminder of its turbulent medieval past.

Often known as Peel towers, they were built between 1200 and 1700, both as refuges in villages and hamlets far from a larger castle, and as places from which warnings could be sounded in times of trouble. Sometimes they were built in lines, so that warnings could be spread rapidly, from one to another, by flags and beacons. A line of Peel towers was built through the Tweed valley in the fifteenth century as a response to raiding parties from the Scottish Marches. The best known of these raiders were the so-called Border reivers (pronounced 'reevers'), who operated on both sides of the Border. They would extract protection money from their potential victims, mostly small landowners. These fees were known as 'black mal'; and by 1552, 'blackmail'. 'Mal' was a Norse word for agreement.[4]

One – quite late – example of a Peel tower is Barnes Tower, on the Tweed, a short distance upstream of Peebles. Peebles was one of the principal towns of the central Borders. It was served by the roads that followed the river from Berwick and Coldstream to Edinburgh. The tower was built in the late sixteenth century as part of an earlier chain that ran the length of the upper reaches of the Tweed. Its outward appearance was severe: a square tower with roughly plastered walls and small windows, with no hint of ornamentation or decoration. The main hall was on the first floor, with the ground floor reserved for livestock, a dairy, food storage and the practical aspects of feeding people. A very short distance away from the tower is the charming mid-Georgian house that was built in the more peaceful times that followed the 1707 Acts of Union between England and Scotland. It is a small and very elegant country house, built in 1773, whose park and gardens grace the gentle slopes leading down to the river. It provides a wonderful contrast with the brooding tower, which still seems to lurk rather furtively within the trees, just a little distance away. But make no mistake, those plain tower houses were not built for appearances. They were all about saving lives, both human and animal.

Raids by reivers were one thing, but sometimes the cross-border attacks were mounted on a larger, more serious, scale. The devastation caused by such an organized cross-border attack was brought very vividly home to me when I directed excavations for *Time Team* in Edinburgh, in the gardens of the Palace of Holyroodhouse. We were looking for evidence for the earliest building of what is now the Royal Palace when, to our great surprise, we came across scarlet-orange layers within the make-up of a palace wall. There could be no mistake: this was clear evidence for a serious fire, which pottery and other evidence suggested had happened in the early to mid-sixteenth century. Of course, you cannot automatically assume that evidence for burning is necessarily linked to an act of violence – bakers' ovens can catch fire at any time – but there was much to suggest that this particular burning was a result of an English army's attack on Edinburgh, in 1544, led by the Earl of Hertford. This attack was part of the mid-sixteenth-century war between England and Scotland that followed shortly after the birth of Mary Queen of Scots in 1542. The birth of a Scottish heir would have been an ideal opportunity for Henry VIII to have united the kingdoms of Scotland and England by a diplomatic marriage. But with Mary's father, the Scottish King James V, dead, Henry tried instead to conquer Scotland in a massive invasion. Known as the 'Rough Wooing', Henry's aim was to destroy the 'old alliance' between the Scots and Catholic France, but he failed. As was the way of the time, Mary was taken to France to marry the dauphin, just four years later, in 1548.

I found the exposure and excavation of those fire-reddened stones strangely moving and actually did most of the work on them myself, despite frequent calls to appear on-camera. There's something fascinating about discovering what looks like a potentially direct link to a known historical event. The next day was the end of the shoot and I decided to drive back home, through the Borders. I stopped at an ancient-looking inn in Peebles, where I enjoyed lunch with some very convivial people – all complete strangers. After the meal, as I was heading back towards the hire car, it slowly dawned on me that the darkness that had coloured my appreciation of the Borders had lifted. I felt far more relaxed. Maybe it was the lesson I had learned from those fired stones: that we English had much to answer for. I honestly don't know what it was – perhaps simply the generosity and warmth of the people. But I realized I could now enjoy this landscape from the north, as well as the south. And that I loved both views.

10

The Boston Stump

A Fenman's Finger of Defiance

The outline of England's eastern coast is broken mid-way by the Wash, a huge square bay some fourteen miles wide at its mouth. The Wash separates the bulging coastline of Norfolk and East Anglia from the gentle curve of Lincolnshire and, north of it, Yorkshire. The Wash ports, principally Boston to the north and King's Lynn to the south, were important trading centres in the Middle Ages, with good access to the wool and textiles produced in Norfolk, the midlands and Lincolnshire. But the Wash has always been very shallow and difficult to navigate, so it was essential that its ports were situated along river channels that were kept open by substantial flows of water. Boston was positioned on the shores of the River Witham, whose deeper channel could be followed across the bay and out into the open waters of the North Sea.

The Wash forms the seaward extension of the Fens, Britain's largest natural wetland. The undulating landscapes around the Fens provided lush grazing for the sheep whose fleeces were the basis of the area's great prosperity in the Middle Ages. Evidence for that medieval prosperity can be seen in the region's many magnificent churches, which are among the finest in Britain, and include the great cathedrals of Ely, Peterborough and Lincoln.[1]

Today, Boston is an unpretentious, medium-sized market town. It still retains a port although it's much further out to sea than its medieval counterpart, which is closer to the town centre. The port area has always been more flexible than the town itself and has been able to follow and adapt to the evolving shape of the River Witham, whose banks were constantly being altered by deposits of sand and silt. Like its counterpart along the Wash shore at King's Lynn, in

Norfolk, Boston has suffered much in recent years, largely due to insensitive local authorities and, yes, low self-esteem. In the post-war decades, planning authorities placed the short-term recovery of the local economy above all else, including the town's unique medieval heritage.

Both towns were immensely important in the Middle Ages, partly due to the trading activities of the Hanseatic League, which can be seen as a north European medieval precursor to the much later European Economic Community. The principal goods exported were farm products, such as wool and grain – and of these, wool was by far the most important. While London has been Britain's principal port since Roman times, Boston came very close indeed to it in the thirteenth and fourteenth centuries, and was still very wealthy in the fifteenth.[2] Decline set in in the early 1500s and by the end of that century the port was virtually abandoned. The port began to expand again in the eighteenth century, but more slowly, a process that continued in Victorian and modern times. Today, grain and other horticultural and arable products have completely taken over from wool as the port's main export.

Immediately inland from the Wash shoreline, dead-straight dykes, rivers and roads break up the flatness of the Fens. In fact, the landscape undulates more than we tend to think. From around 500 BC onwards, especially in the area around the Wash, tidal bars and silts would build up, often after storms. Slowly, land levels began to rise; from Roman times, the land surrounding the Wash was dry enough for people to settle there. These settlements continued to expand during the Middle Ages. This essentially piecemeal settlement pattern gave rise to a landscape characterized by twisting lanes and meandering watercourses. The higher ground around the Wash is known today as Marshland. Fine medieval churches, built on the coarser-textured marine deposits, are an unexpected feature. And of these, by far and away the most spectacular is the Stump, which can be seen across the Fens to the west for at least fifteen miles.

The constant danger of over-simplifying history is well illustrated by the story of the Boston Stump. It has become quite fashionable to declare that the tower (the Stump) of St Botolph's is too tall; that it is out of proportion with the rest of the church. Personally, I have never found that. To see St Botolph's at its best, I suggest you cross the River Witham a short distance downstream, and on your walk call in at a pub and enjoy a glass (or two) of Bateman's excellent real ale. Then return towards the church, but on the far side of the river. Walk past the huge west window for about a hundred yards, then turn round and look back. What you will see then is surely perfection, in both scale and proportion. But you have to stand back to appreciate it.

It has been suggested, with some justification, that the Stump's builders' original intention was to erect a spire above the two lowest storeys of windows but that plan had to be abandoned when it was realized that the ground would not bear such a load. Certainly a spire would fit in well with local practice. Lincolnshire is famous for its spires and some of the finest in England can be see on parish churches in Louth, Long Sutton and Grantham. Tall and elegant, they often feature pinnacles and slender flying buttresses, which were added for support but which seem to launch them into the clouds.

The great architectural historian Nikolaus Pevsner, who seems to have taken particular delight in the churches around the Wash, is of the opinion that the two uppermost storeys of the Stump were built out of pride, or hubris, and a dogged determination to display the great wealth that Boston enjoyed.

The building of the tower took place after the main body of the church had been completed. Work on the tower began around 1430 and had been finished less than a century later – remarkable progress, given what the builders and their families had to contend with. These were decades of successive waves of plague, following the initial impact of the Black Death in 1348.[3] Yet, throughout epidemics that decimated England's population, the building of the tower continued, ever higher. I have lived in the Fens for almost fifty years now and I sometimes wonder whether there wasn't a sort of Fen logic behind what they did. You could construe the Stump as a tribute to their faith, which it most certainly is. But I think that was only part of the story. I have come firmly to believe that they also saw the Stump – their Stump – as a sign of defiance: a symbolic Fenman's finger, making a rude gesture at approaching Death.

II

Romney Marsh

Remote, but not Entirely Forgotten

In the past, historians were inclined to believe that the diversity of Britain's geology and geography played the major role in shaping the country's landscape; the human response to these two natural factors resulted in the character of particular landscapes. This human response, they felt, was a largely predictable, almost mechanical process, which in turn served to heighten the perceived importance of the underlying natural factors in the landscape's evolution. Ideas of this sort invariably acquire labels: this view of landscape formation was described as 'environmental determinism'. I never really got to grips with the pros and cons of it, largely because I couldn't face becoming embroiled in an academic debate, but also because, as somebody who also had a life in a real landscape outside the groves of academe, I could see the arguments were largely irrelevant: of course people adapted to their surroundings. They had to. But they did so in very different and often unpredictable ways. You can see this when you look more closely at two geographically widely separated, but superficially similar, flat, wetland landscapes: the Fens (where we have just been) and Romney Marsh, on the south coast, in Kent.

I have spent most of my life in the Fens, and I was astonished at the differences I found when I began to look into the history of Romney Marsh. The region is very much smaller than both the Fens and the Somerset Levels, England's largest and second largest wetlands. Of course there are obvious similarities in all three, which are mostly to do with the control of water. All flood-prone landscapes would be uninhabitable if water was not removed as quickly as possible; in the past, this was done using wide, straightened rivers and drainage channels. Similarly, tidal floods and storms originating out at sea have to be prevented by banks, walls and other coastal defences.

Romney Marsh occupies about a hundred square miles on the coast of south-eastern Kent, about ten miles south of Ashford. It is bounded to the north by the Royal Military Canal, which is twenty-eight miles long and follows the line of a low escarpment that forms the edge of the Marsh between Seabrook, near Folkestone (in the east) and Winchelsea (in the west). It was built between 1805 and 1809 at first by civil, then by military engineers, as a defensive measure against an expected invasion by Napoleon's forces. That invasion never materialized, but the canal survived, to be used as a barrier against smugglers from the Marsh (of whom Russell Thorndike's smuggler Dr Syn is the most famous fictional example). By 1877 the canal had been partly abandoned, but at the start of the Second World War it again came into use as a line of defence against German invasion. This involved the construction of numerous concrete pillboxes, which still survive. Today a footpath runs along the entire length of the canal, which has become a major nature reserve and visitor attraction.[1]

So did the landscape of Romney Marsh somehow predetermine the existence of the Military Canal? When I first learned about it I was immediately reminded of two broadly similar features in the Fens: the Car Dyke and the Ouse Relief Channel. The Car Dyke runs north–south along the edge of the Fens in south Lincolnshire and in Cambridgeshire, especially around Peterborough.[2] This great ditch and bank was dug and maintained in Roman times, when it probably formed part of a larger canal system linking the Fens with the Humber. There is also evidence to suggest that in places it would not have worked as a practical canal and was dug instead as a drain to catch floodwater running off higher ground. In the southern Fens, around the River Ouse, flooding incidents in the late 1930s, followed by disastrous inundations in 1947 and 1953, led to the construction of a cut-off channel to catch water from higher ground and take it directly towards the Ouse outfall.[3] From the air, the Military Canal, the Car Dyke and the recent Ouse Relief Channel all look very similar, but their histories could not be more different. So did the landscape 'determine' how people behaved? My answer is a firm No. The form

of the landscape may have affected the shape and positioning of the
various canals, dykes, banks and ditches, but their creation was the
result of human will and inclination alone – which in turn was
shaped by the entirely unpredictable course of history. It is also
worth noting that there are many places, along, for example, the
Norfolk fen-edge or the Witham valley in the Lincolnshire Fens,
where catch water drains would have worked well, but were never
constructed.

The coast of Romney Marsh forms an open V-shape, with the
promontory of Dungeness at its centre. To the north and east it runs
through Dymchurch to Hythe. This stretch of beach includes four
Martello towers, which were built at the same time as the canal, to
house artillery – a first line of defence against a seaborne invasion.
The coastline from Dungeness to Rye, in the west, is more low-lying
and forms the edge of Walland Marsh to the north. Walland Marsh
was drained and reclaimed in the later Middle Ages, slightly later
than Romney Marsh alongside it, to the north and east. The two
marshes (which I treat here as one) are separated by the thirteenth-
century Rhee Wall, a bank that was part of early attempts to drain
silts from out of New Romney harbour.

As a Fenman, used to vast open vistas with occasional glimpses of
distant hills, I find the more intimate scale of the Romney Marshes
immensely appealing.[4] It used to be believed that 'remote' landscapes
such as high moors and low-lying marshes had not been occupied in
early times. Today we can appreciate that this was very rarely the
case. And Romney Marsh is no exception. We now know the area
was occupied in prehistoric times, and I suspect we have probably
underestimated the size and permanence of those communities, just
as we have done in the Fens. Hints of the region's pre-Roman life can
be found in the recent discovery of a sea-going vessel at Dover and a
timber causeway that resembled Flag Fen, in marshes at Shinewater
Park, near Eastbourne.[5] In Roman times, the wet grasslands were
grazed by sheep in summer, when ground conditions were suitable,
while salt was extracted from seawater trapped in inland lagoons.
But what most took me by surprise, on my first visit to Romney
Marsh, was the form and appearance of the numerous fine churches.
They were so varied and appealing, yet so unlike the towering stone

buildings that the prosperous medieval wool trade had bequeathed to the Fens. There are many probable reasons for this: fewer larger abbeys, with their well-run estates, together with smaller parishes and landlords whose principal holdings lay outside the Marsh.

In Saxon times, areas of the marshes had become habitable and during the early Middle Ages many of the drained salt marshes had become permanent pasture. Farmers in the area had even developed their own breed of sheep – the Romney Marsh. They are quite large animals, with good fleeces and a natural resistance to foot problems caused by damp ground.

In the early Middle Ages, the population of the Marsh grew steadily. Mixed farming – crops and livestock – prospered, and the villages grew in size. Soon Romney Marsh became the most heavily populated district in Kent. Then in the 1230s the area was hit by some serious storms, which culminated in the great storm of 1287 which breached the shingle barrier, and the sea rushed in. Romney harbour was filled with silt and never fully recovered. To make matters worse, in the fourteenth century the region was ravaged by plague; by the end of the century its population had halved. As elsewhere in rural England, the smaller population led to labour shortages on the land. So farmers, and landowners (especially the larger ones), took to the raising of sheep in huge quantities. The raising of sheep requires far less labour than arable farming: fields do not have to be ploughed or harrowed and the cutting of hay for winter forage is far less arduous than reaping, storing and then threshing grain. The sheep trade continued to be the mainstay of the Marsh economy right through to the nineteenth century, when the Romney Marsh breed was exported to Australia in considerable numbers. But the sheep trade did not require a large labour force, and from being the most heavily settled part of Kent in the thirteenth century, by the mid-seventeenth century it had become the county's least populated region – a situation that persisted into Victorian times.

Such sharply fluctuating fortunes have left their mark on the parish churches of the region, which are strikingly diverse in size, shape and date. Some interiors preserve unusual seventeenth- and eighteenth-century box-pews, which have survived because the parishes were never large nor prosperous enough to replace them with the heavy and rather characterless Victorian benches that are such a common feature in so many urban and rural churches elsewhere in Britain. I also have a particularly soft spot for the Romney Marsh churches because they are wonderful examples of modern conservation. In the 1980s it was realized that the churches of the Marsh were in poor condition and the small parishes often lacked the funds to maintain them. So in 1982 the Romney Marsh Historic Churches Trust was set up. Soon there were articles and appeals in the national press and a successful fund-raising campaign was mounted. Thanks to the Trust, and the people behind the individual parish churches, the churches of Romney Marsh are now recognized, collectively, as a unique national treasure. With their hanging tiles, quirky towers and higgledy-piggledy graveyards, the churches of Romney Marsh are such a contrast to the soaring magnificence of the great Decorated and Perpendicular churches that the later medieval wool trade bequeathed to the Fens and eastern England. Their interiors too are a delight: as rich and varied as their exteriors. My personal favourite is the smallest, loneliest, most remote church in the Marsh.

The church of St Thomas Becket, Fairfield sits isolated on a low man-made mound in a marshy field. When it was built in the late twelfth century it was surrounded by the houses of a small village, which has long since vanished, to be replaced by bare fields of sheep and grass. The steep roofs were originally thatched with the reeds that grow so luxuriantly in the surrounding landscape. I had expected this well-known church to be somehow lonely and forgotten. But to my surprise that wasn't what I found when we visited. Far from it; the place felt loved.

I have to confess that I have never been a purist when it comes to buildings. It doesn't concern me that large parts of St Thomas Becket have been rebuilt. It's part of a natural process, a continuing tradition; timber buildings in wet areas are always more liable to decay and the restoration work, by the architect W. D. Caröe in 1913, was at pains to re-use fifteenth-century timbers and eighteenth-century bricks, which had themselves replaced the wattle-and-daub of the original, late twelfth-century walls. So, all in all it's a bit of a dog's breakfast – but a perfect one. It represents what happens when people care about, when they love, a building. Inside, in stark contrast to the open, windswept landscape outside, a sense of warmth exudes from the timber roof and low, massive beams that frame a superb set of painted eighteenth-century box-pews in the nave. The village outside it may have vanished, but I don't think I have ever come across a church building that has been more cherished.

12

Medieval Productivity

The Open Fields of Laxton, Nottinghamshire

I like working landscapes best of all. Most have their roots firmly in the past, but their early history lies deeply concealed in the shape, layout and dates of farms, barns, churches and fields. Very few places indeed retain the working practices of the Middle Ages into the present day. One of these is the Nottinghamshire village of Laxton.

I first learned about Laxton in history lessons at school. Through various accidents of history it had become the only surviving example of an Open Field parish still operating in England.[1] The modern pattern of agriculture in England began in the early 1700s with the so-called Enclosure Movement. Enclosure took place parish by parish and was usually overseen and organized by the local lord of the manor. It gave rise to the pattern of fields, lanes and farms that can still be seen in rural areas today. Laxton, however, lost its lord of the manor in the seventeenth century and the various farmers in the parish didn't feel the need to enclose the land to form individual farms. The old system worked well for them, so they retained it. By 1952 the village's unique status was recognized, and some 2,000 acres were purchased by the Crown Estates to be protected and farmed in the traditional way.

At first glance, Laxton looks much like any other east midlands village. It lies in the low hills on the western side of the Trent valley, not far off the Great North Road (the A1), some seven miles northeast of Newark-on-Trent. On my first visit there, in the 1990s, I was greeted by a typically English rural scene: a gently undulating landscape with large arable fields, interspersed with pasture for sheep and cattle. Here and there were trees and woodland, with hawthorn hedges and green lanes. As we approached the village I could see a glint off recently ploughed soil, a reflective shine that told me instantly this was heavy clay land. I was also aware that the pre-Enclosure, medieval system of farming, known as the Open Field System, worked particularly well on heavy clay land: doubtless one of the reasons why the men of Laxton didn't want to change their traditional ways. One tends to think that 'traditional' is often another way of saying 'simple' or 'straightforward', but not in this case. So first, a few words of explanation.

During the Middle Ages, roughly from just before the Norman Conquest right through to the seventeenth and eighteenth centuries, English rural life was based around the manorial system. Each village was run by a manor, which had its own court, supervised by the lord of the manor, or an appointee. This court decided on all disputes over rent, farming and similar matters. The land was owned by the lord of the manor, who rented it out to the villagers in return for their labour, or even their military service. For their part, the peasant inhabitants of the village were allocated enough land to feed their families, and maybe to make a small surplus in good years.

At Laxton the land to be farmed was originally laid out in four large Open Fields and each tenant was allocated strips of land within them. Everyone was expected to farm the strips in each Open Field in the same way. One field would be left fallow, to regain fertility, while the other three would be given over to spring or winter-sown crops. Somewhat later one of the fields was dropped, so today Laxton possesses three large Open Fields, which still retain their medieval size and shape, although the tiny strips of the medieval farmers are now much larger, to suit modern tractors and equipment. Together the three types of cultivation – fallow, autumn- and spring-sown crop – formed what is known as a three-course rotation. It was a system that encouraged soil growth and fertility, but discouraged the development of diseases. Variants of it are still in use around the world.

The Open Field System was most extensively employed across the central zone of England, where the land was generally heavy but fertile. Though Open Fields existed elsewhere, the system was less rigidly enforced and livestock and woodland played far more important roles. Woodland was seen as yet another resource, which was

carefully managed to provide renewable supplies of timber for building, logs for fires and flexible wattle rods of hazel, which were woven
into house walls, hurdles and fences.

I was already aware that Laxton had a small castle, so I thought
that would be the best place to start. It was reached down a short
green lane, lined on both sides by trees and hedges. Incidentally, I
couldn't help noticing that many of the trees were ash, which loves
the heavy clay soils of the midlands. Sadly, these are all under threat
from the fungal disease ash dieback, which probably couldn't have
spread so far and so fast in the Middle Ages.

The castle is quite unlike the spectacular towered structures of
popular imagination. It was my kind of castle. Not exceptional: I
suppose you could see it as a 'jobbing' castle, a place that existed to
project the power of the ruling authority. In its way, the medieval
equivalent of the substantial Georgian country houses that are still a
feature of the rural midlands. Like many rural castles built by the
incoming Norman aristocracy in the decades after 1066, its strength
consisted of steep banks and deep ditches, with a wooden or stone
tower, known as the keep, which sat on top of a central mound, or
motte. The aim of the incoming aristocracy was to establish their
power and retain it, and their castles were built to do just that.

The castle at Laxton had been carefully placed at the top of a
steep escarpment in a naturally strong defensive position. The layout
of the castle is remarkably well preserved, beneath a thick covering
of grass. The earthworks date to early Norman times and consist of
a high motte and two banked, defended yards, known as the inner
and the outer bailey. The lord's family and retainers lived in the timber keep, atop the motte, and in times of trouble the inner bailey
protected the villagers, whose livestock were kept safe in the outer
bailey, both of which survive almost intact. Some masonry walls,
built in the thirteenth century, survive on the banks of the baileys.
The unpaved green lane I had followed from the village branched left
just before the castle and then ran straight for a few hundred yards,
out to one of the three surviving Open Fields, the West Field. This
grassy, ash-tree-lined lane, which runs along and behind the village,
is called Hall Lane.

Hall Lane captures the essence of what it would have been like to have lived here in the Middle Ages. Each of the houses that faced on to the main village street had long rectangular back gardens, known as tofts, which were much larger than modern urban gardens: they had to provide families with milk (from a house cow), pork, eggs, fruit and vegetables. Hall Lane ran along the back of the tofts and would have allowed people, animals and produce to move freely between different properties. It also gave access to the castle and the Open Fields. Today it is a wonderfully peaceful hedged and tree-lined green lane, but in medieval times it would have been bustling with activity. What a tractor can today achieve in a few hours would have taken days. So Hall Lane would normally have been busy, with men and teams of horses and oxen constantly passing along it. The lane is fully wide enough for two modern trucks to pass each other, with room to spare, and this shows it had originally been well used. On heavy clay land in wet weather, carts and other vehicles need to avoid deep waterlogged ruts, which is easier to do if the roadway is sufficiently wide.

As you stroll down Hall Lane today, you can quite clearly see the boundaries between the tofts, and just as in medieval times, each holding is different: mown grass and miniature goal posts for children, vegetable patches, flower beds and chicken sheds. Those chickens, flowers and vegetables told me that the cottages in Laxton were still lived in by real country people, who didn't buy all their food from supermarkets.

Eventually I reached the end of Hall Lane, which simply stopped. Ahead of me stretched a vast plain of stubble. It wasn't remotely what I'd envisaged. It resembled nothing more than a modern 'grain plain', of a type I am only too familiar with in the Fens. So my mental picture of the medieval countryside had been wrong. I don't know why, but the sheer wide-openness of the Open Fields had escaped me. I'd somehow imagined there'd be more trees and hedges, but of course that would have been ridiculous in such an intensively ploughed and cultivated landscape. This is because the teams of oxen and horses needed to pull ploughs through the heavy clay soil could only be turned around at the end of each row if there was adequate space. In such landscapes, trees and hedges simply got in the way.

The rural landscape at Laxton is one of my favourite places in

England, because it replaces our sometimes rather romantic view of courtly medieval life with some gritty realities. Before the Black Death of 1348, the population had been steadily increasing, and in some areas there was clearly pressure on food. So the rural economy had to rise to the challenge, and this is what Laxton is essentially all about. As a way of productively farming heavy clay land, the essentially collective and co-operative Open Field System worked very well. It meant that the village's farmers had access to the lord's ploughs and teams of oxen and horses, as part of their tenancy agreements – and without such practical measures cultivating the heavy clay soils would have been impossible. Co-operation would have required agreement among the villagers, and numerous disputes about who had what and when would have arisen almost daily. Most of these would have been sorted out by the lord's agent, but sometimes a higher authority, the manorial court, was needed.

The early medieval manorial court would have convened in the castle, with the lord of the manor in the chair. It was a form of rapid justice, where one man was in control, but ordinary villagers had a strong voice. And it was an open system, where cases were made and judgments were given in public. Ultimately, it was in everybody's interest – both the lord and the commoners – that the administration of justice helped to maintain the local economy and allowed it to run smoothly. Disruption would have favoured nobody.

We have long known that the Open Field System worked effectively and helped to maintain a reliable supply of food to a large and growing population until the start of the fourteenth century, when practical problems, such as poor harvests, were encountered. Some fifty years later, the Black Death and successive waves of plague were to provide a lasting, if very cruel, solution to these difficulties. But we have tended to under-estimate the sheer efficiency and effectiveness of the earlier system, which may have been based around individual manors and parishes, but which, we now realize, depended upon sophisticated internal and external supply chains. Powerful horses – what today we would call carthorses – were the heavy movers of the age and they were bred and traded through a complex and sophisticated system that extended across the country.[2] They hauled the large loads of grain that were needed to supply urban markets.

Medieval prairie farming: the West Open Field at Laxton, Notts.

Manure was the medieval equivalent of modern chemical fertilizers, and the manorial system ensured that all the peasant holdings within a parish received their supply, which allowed the Open Fields to produce far more than they would have if left to nature alone.[3] So those lovely tree-lined leafy lanes at Laxton have suddenly acquired a new relevance: they were once part and parcel of a highly productive working landscape, which, I'm glad to say, continues to this day. This is no re-enactment. It is real, living history.

13

Ironbridge

Where It All Started

The Severn Gorge near Telford in Shropshire has always been a daunting prospect to cross. The river itself flows fast and turbulent, and is quick to rise when there is rainfall in the hills to the west. The energy of the water has cut a deep, steep-sided rocky gorge that proved to be a major barrier to travellers heading east towards the emerging new industrial centres around Coalbrookdale and modern Telford. And then, in the 1770s, what must have seemed like a miracle happened: the world's first massive iron bridge was constructed across the gorge. It was built without fuss and disasters and is still standing: a tribute to the genius of those early ironmasters and the skill of the men who worked with them. For me, Ironbridge is the Stonehenge of the industrial era. Every time I walk across it, I feel incredibly privileged. Indeed, it's probably as close as I will ever get to a religious experience.

The bridge, built nearly 240 years ago, is immensely impressive. Its great central span springs off arched piers on either side of the gorge, which were themselves rebuilt about forty years later when their timber precursors were in need of repair. This detail put me in mind of the timber frame within the great spire at Salisbury Cathedral: wood with stone. In medieval times, builders were far more relaxed about mixing wood with other more durable materials than was the case by even the mid-nineteenth century. Builders were prepared to return to their projects and gradually make them more permanent. People were less obsessed with instant perfection: it was recognized that some things took time to get right.

I tend to approach Ironbridge with respect and a degree of deference, much as a religious person might approach the high altar in a great cathedral. I still find that the best way to appreciate the magnificence of the structure is from below. Walking slowly along the river bank beneath, I wait until I'm directly under the centre of the bridge before allowing myself to look up. It's an awe-inspiring experience, and I must have done it half a dozen times. The hundreds of iron girders seem to float upwards, with extraordinary lightness. The bridge itself feels organic, as if just flexing its muscles after centuries of repose.

When first looking at the bridge this way, I remember being struck by the lack of bolts and rivets. At the time I had been researching ancient woodworking, and I immediately recognized that the individual girders and spars high above me had been joined together using carpenters' joints, which were integral to the pieces concerned. In other words, each piece had been shaped to join with those alongside it. And it would have made excellent sense at the time, because many of these joints get stronger when under compression. The techniques those early ironwork engineers used had been borrowed from the more ancient traditions of carpentry, just as the prehistoric builders of Stonehenge had used mortice and tenon joints to join the massive uprights and lintels.

I then walk further along the gorge. This is another world. We could be a hundred miles away from the bustling heart of nearby Telford: trees and shrubs perch perilously on the craggy sides of the gorge; herons stand grey, still and alert, staring intently down into the waters. Then the path turns and I climb back up to the town level, where I make my way to the bridge for the culmination of my visit: a slow walk across, and then back, high above the gorge. Sometimes I might stop for a bite of lunch on the other side, but not until I have crossed the bridge in both directions, pausing to gaze downstream when I reach the centre, on my outwards walk towards Wales, and again, this time looking upstream, on my return. I know it sounds obsessive, but that's how I like to do it – and it ensures I miss nothing. I also like to admire the unostentatious toll buildings on the west side of the bridge, which prominently display a list of charges for people, horses, carts and wagons. These businesslike, quietly elegant buildings are typical of those you would have encountered on many a turnpike road in their late eighteenth- and early nineteenth-century heyday.

I have avoided using the phrase 'Industrial Revolution'. A true revolution is something that happens quickly. The development of industry in Britain, however, spanned four hundred years: beginning in the sixteenth century, it was still under way in the first half of the nineteenth.[1] This idea of it being a revolution was, however, underscored by the prominent role played by certain individuals, often known as ironmasters, of whom the Darbys of Coalbrookdale are among the best known. Traditionally, such men are portrayed as lone geniuses, the entrepreneurial equivalent of Newton, whose ideas blazed a pioneering trail. We know now, however, that this was far from the case. Rather, they developed their new ideas in close collaboration with the skilled and very knowledgeable men who worked with them in their forges and workshops – or, to put it in modern terms, the early ironmasters were highly gifted and creative team leaders.[2]

The teams they led would not have been possible in a country with a more rigid social, political and religious hierarchy. Britain had gone through some major changes at the end of the medieval period and, if anything, these accelerated after the English Civil War of the mid-seventeenth century. What we saw was a huge loosening of the medieval system of inherited class distinction, which in rural areas was based around the manor. This is the system that still survives, albeit altered, at Laxton. The process was hastened by three centuries of epidemics, starting with the Black Death of 1348 and lasting through to London's great plague of 1665–6. Continuing shortages of manpower gave greater influence to the workforce and accelerated change. Although, as we will see, further reforms were still to come, by the start of the eighteenth century Britain was becoming an excellent place to be an entrepreneur.

The landscapes of the upper and mid-Severn valley, though, had never been part of the manor-based Open Field System so characteristic of the English midlands. Instead, a less regulated system had developed, based around livestock, grazing and woodland. By the 1600s we see the rise of a new class, the so-called 'yeoman farmer', who owned his own land and was not directly answerable to any lord of the manor. This new class of independent smaller farmers could not rely on rents or other sources of income to feed themselves and their families over the leaner months of the farming year, which usually meant winter and early spring. So they were prepared to turn their hands to almost anything to make a little money. Some people diversified the wool side of their sheep enterprises, turning to spinning and weaving. Others tried their hands at mining and mineral extraction. Still others looked to smithing and metal-working. By the middle of the eighteenth century, many of these once-secondary sources of family income had grown in importance, so the spinning and weaving that had once taken place at home was now being done in workshops and even in factories. The metal-working that had begun around Coalbrookdale was now prospering, and workshops were proliferating.

The arrangement of settlements along the upper Severn valley around Coalbrookdale is rather different from what you might find elsewhere in England. The familiar hierarchy of hamlets, villages and small towns is replaced by a more dispersed and less readily defined pattern, which today resembles light urban sprawl. If you carefully examine the date of the various buildings, it becomes possible to identify what would once have been separated farmsteads and workshops. This reveals a pattern of settlement that did not focus on, or cluster around, a major town or city. Each unit was self-contained and independent. We now realize that this was because they needed plenty of space around them to store ore and fuel, and – just as important – to allow smoke and fumes to disperse.

The great bridge at Ironbridge was built between 1777 and 1782, but it could not have happened without what may well have been the biggest single technological development of the early industrial era.[3] In 1709 Abraham Darby I worked out how to smelt (i.e. extract) iron from iron ore by using coke, a form of pre-heated coal. Before that, the smelting process had had to use wood charcoal, which was both slow and relatively expensive to produce. The iron for the great bridge was manufactured by Abraham Darby III in Coalbrookdale, and the bridge itself was designed by the wonderfully named Thomas Farnolls Pritchard. If you are not just passing through, I would suggest you spend the rest of your day at Coalbrookdale in the Museum and walking round the Darby furnaces and forge. I can't think of anywhere better to steep yourself in the pioneering phase of the modern world.

The landscapes of the early industrial era are unsung monuments to those four post-medieval centuries of social and political change. Today schoolchildren learn about kings, queens, politicians and great generals, but for me, men like Abraham Darby in far off Coalbrookdale were just as remarkable. They were the first great technologists, who were able to alter the world around them through their own abilities, rather than their birth and inheritance alone. I find places like Ironbridge uplifting: they project the optimism, the guts and the sheer, bloody-minded determination that ultimately gave us the modern world. The town is named after the great Iron Bridge, but there is far more to it than that. Many of the workshops still survive along the gorge and in Coalbrookdale. Yes, they have been restored and made visitor-friendly, but the heat-reddened stonework of the furnaces tells its own story of the hard, dangerous, yet skilled work that took place there. It remains a living and constantly developing story, because every year local industrial archaeologists are producing fresh revelations. And the more we discover, the more we respect those extraordinary self-made men and women who did so much that affects our lives to this day.

14
Buckler's Hard
Yesterday's Technology

The south of England is becoming increasingly overcrowded, but occasionally one comes across places where the openness and scale of the landscape make you forget that this is the case. Buckler's Hard, a long-abandoned shipbuilding village on the River Beaulieu, a small river draining into the Solent, on the Hampshire coast, is just such a place. Incidentally, a 'hard' is the local name given to a gently sloping beach, or shore, where boats can be pulled out of the water, or launched. A small but sprawling hamlet, Buckler's Hard commands views across the wide expanses of the tidal river and south towards the North Solent Nature Reserve. It's an undulating landscape of sand dunes, saltmarshes, scrub and wind-gnarled trees. It was an important, bustling place in the eighteenth century, but it can teach us real lessons for our own times.

Two hundred and fifty years ago, Buckler's Hard was an important shipbuilding community. It was, and remains, part of the Beaulieu estate, owned by the Dukes of Montagu. Originally known as Montagu Town, the hamlet was first established in the early eighteenth century by the second Duke of Montagu, who had ambitions to turn it into a major port to trade with the then rapidly growing sugar-based economy of the West Indies. Soon, though, its leading industry became shipbuilding. The shallow, gently sloping river banks were ideal locations for building and launching wooden vessels, and the nearby New Forest provided abundant supplies of good quality timber, especially oak. In 1744 a new master shipwright, Henry Adams, was appointed and soon Buckler's Hard was building vessels for the Royal Navy: three of its ships eventually took part in the Battle of Trafalgar. However, as timber ships gave way to iron during the nineteenth century, the yard slipped into decline, and by the end of the century, it had to close.

The survival of Buckler's Hard owes much to an accident of history: it was always part of a large estate. If it hadn't been, when the yard closed it might have been adapted, and built over; or perhaps turned into a small industrial centre, providing ships, supplies and accessories for the Navy, the fishing trade, or merchant shipping. But this never happened; instead, it became a sleepy residential village on a great estate. Things were to change dramatically, however, during the Second World War. Once again, Buckler's Hard played its part in great events, with the construction of the new, lightweight and highly effective motor torpedo boats. Then, in early 1944, the river became home to hundreds of landing craft, which were concealed and moored there in the run-up to D-Day.

The houses of Buckler's Hard were used to accommodate estate staff, offices for administration and even a chapel. But they were never substantially altered and still retain their early Georgian rural charm. They line what is essentially a very wide street that runs down the slope of the hard, to the water's edge. The two terraces stop well short of the high-tide line and are fronted by buildings that overlook the river. Today one of them, the Master Boatbuilder's House, is a hotel.

Very often the most heavily used surfaces of hards were reinforced by gravel or paving, which happened here, but in a restrained way. The buildings on either side of Buckler's Hard were constructed at a time when people seemed incapable of making anything that was ugly. They are elegant and unpretentious two-storey houses with dormer windows in the attics. These are not urban terraces: their informality and lack of pretension are far more rural; the brickwork of their walls (made from local clays) calls for roses, not graffiti.

Modern films often portray industry in the eighteenth and nineteenth centuries as a class-based system, where remote and often uncaring bosses, in suits and top-hats, barked orders to a grimy, down-trodden workforce. Although that may have been the case sometimes, it was by no means the rule. We saw in the previous chapter how the Darbys of Coalbrookdale developed their ground-breaking new techniques in conjunction with the skilled craftsmen they employed, and something similar was happening in shipbuilding, where the master shipbuilders worked alongside their workforce. This sense of collaboration is captured perfectly by the buildings at Buckler's Hard.

The Master Boatbuilder's House, which fronts the left-hand terrace as you approach the river, is a fine Georgian building, but it it not overwhelmingly grand. It complements, rather than dominates, the other buildings around it. If the buildings of Buckler's Hard were people and part of a team, this house would have been that of the team-leader. One can well imagine the master boatbuilder taking a stroll on a warm summer's evening and mixing with the men and their families who helped to create the great ships that made Britain the world's leading maritime power.

The landscape around the little hamlet is well worth exploring, and there are many good footpaths and walks that lead out into the surrounding woods, dunes and saltmarshes. Living near the Wash, I have long been a fan of such landscapes and the abundant wildlife they attract, but this time I was in for a surprise. Not far from the Hard, but apart from it – its detachment emphasized by the clump of trees surrounding it – is the garden shed of my dreams: the Duke's Bath House. This is a thatched 'cottage ornée' built in 1760 for the third Duke of Montagu's son, Lord Brudenell, as a saltwater bathhouse. At the time, bathing in seawater was believed to have healing and soothing medicinal qualities. In urban areas, saltwater bathing became popular in the early nineteenth century, especially following cholera outbreaks in the 1830s.

Earlier I suggested that Buckler's Hard had lessons for our own time. It was preserved because although technology had moved on, the family that owned the shipyard and the estate did not then demolish all the buildings. Whether it was respect for the old ways of doing things, or simply an unwillingness to tear down perfectly good housing, I do not know, but the fact is that the old shipbuilding village survived and has recently come into its own again, as a visitor attraction (which probably brings in as much income as the original boatyard). I think today we are far too keen to sweep away the 'unsightly' remains of industries that are no longer wanted. The survival of Buckler's Hard should teach us that we should pause before we rush to develop all the so-called 'brownfield' sites, where successive governments would like us to build houses for the future. Recent industrial history may seem outdated and irrelevant now, but will our great-grandchildren necessarily think the same?

15

The Quiet Revolution

The First Turnpike at Caxton, Cambridgeshire

When I was young, my family would pay regular trips to Cambridge, to visit my uncle's family. We lived in a small village in the gently undulating chalk hills near the small town of Baldock in north Hertfordshire. In the eighteenth century my Quaker ancestors built a brewery there and the Pryor family have lived in the area ever since. Our route would take us past the south Cambridgeshire village of Caxton, and on its outskirts a weathered gibbet with a vestige of rope still dangling forlornly from it. At which point my nine-year-old imagination would go into overdrive, conjuring up a lurid image of a highwayman's body hanging from it, his distraught wife and family weeping on the ground below. Back then, I didn't realize that the gibbet itself was a later restoration, but it didn't matter. It achieved its aim: ever since, I have loathed the obscenity that is capital punishment.

The straight north–south road we drove along to Caxton was first laid out by the Romans, who named it Ermine Street; later, it became known as the Old North Road. Its present status as the lowly A1198 belies its historical importance in the development of England's infrastructure.

Most accounts of the early industrial era start with the rise and development of those precursors to the railways, the canal networks. These often involved heavy civil engineering projects, such as tunnels and aqueducts. Such projects made the reputations of the great civil engineers: men like John Rennie and Thomas Telford. Roads, by contrast, rarely get mentioned. Nevertheless, they also played an important part in the development of modern industry and, most particularly, in the way it was managed and marketed. Canals were able to move bulk goods, for example grain, or coal, in large quantities and very cheaply, but they were slow. Mail needed to be moved quickly, as did papers, books and lighter products such as clothes, and food also needed to travel faster. Communication then and now was about meeting new people: new clients, new suppliers and new bankers – all of whom were essential to any expanding enterprise. These things required an efficient road system.

Up until the later 1600s, the maintenance of England's road network was the responsibility of local authorities, even in poor rural areas where money was very short. Many roads were in disrepair, and some were abandoned, to become green lanes. Then in 1663 it was recognized that an important arterial route, the old Roman Ermine Street in the counties of Hertfordshire, Huntingdonshire and Cambridgeshire – then in a very poor condition, after decades of neglect – needed substantial repair. Materials to carry out the necessary roadworks had to be brought in, but none of the parishes through which the road ran had the money to pay for the work.[1] So a Turnpike Trust was established, and, like the many others that followed, it had to be approved by Parliament.

The idea of forming Turnpike Trusts had begun in Hertfordshire in the 1650s, when parishes along the Great North Road (now the A1) petitioned Parliament for help in maintaining the heavily used road. Parliament established the Trusts to supervise the spending of public money and its repayment, through tolls. The first trusts were project-specific and looked after short stretches of pre-existing roads. After 1663, turnpikes became substantial, self-financing roads that frequently involved widening and straightening roads, and even the building of entirely new stretches. The term 'turnpike' refers to a turnstile or gate that road users could only pass through after they had paid a fee, which went towards the maintenance and improvement of the road along which they were travelling. Usually, such upkeep involved the clearance of obstacles, unnecessary diversions, and the provision of milestones and signposts. Soon toll-houses, as they were called, developed their own distinctive style of architecture: polygonal, many-windowed structures with clear views of the road in several directions. You can still spot hundreds of them on roadsides today. As the better-maintained roads proved more popular, especially in the eighteenth century, coaching inns sprang up. These inns, often featuring a large central arch leading through to a stable block at the rear, provided fresh horses and secure stabling and, for travellers, food and a bed. In many towns, these new inns became destinations in their own right and many offered rooms where people could set up business meetings. It proved a successful, cost-effective system for providing modern roads, which was soon adapted across the British Empire, and in the United States.

By and large turnpikes were self-financing, and they transformed England's road system. And it all began on Ermine Street, which for me today is England's most atmospheric road.[2] It doesn't pass through spectacular scenery, but runs through eastern England, from the Thames valley, skirting the Chiltern Hills and the western fringes of the Fens before crossing the Lincolnshire and Yorkshire Wolds. As you follow its course, you will soon get to spot the familiar milestones, toll-houses and, of course, those numerous roadside inns. Hundreds, probably thousands, still survive along England's roads and can easily be spotted. I like the scene in Caxton particularly: an ancient village in rolling hill country with woods and high hedges. Somehow it manages to retain its character, despite the burgeoning presence of Cambourne, a spin-off small new town satellite of Cambridge, a mile or two to the east.

When Cambourne was built, the A10 was diverted, thereby removing the queues of traffic that used to choke the centre of Caxton. The two roadside inns, the Crown and the George, near the brow of the hill, on the west side of Caxton, are particularly fine and when seen from the front appear to be eighteenth century. But go round to the back and you will find they are actually Elizabethan.[3] This re-fronting of earlier inn buildings is a distinctive feature of towns and villages along Ermine Street, and many other turnpikes. It shows how successful the road improvements were in encouraging new traffic, and also illustrates the political and economic pressure to form the first Turnpike Trust. These added-on Georgian frontages are all about promoting the inns to the many travellers that journeyed along the upgraded road.

There is something so dull and ordinary about modern road labelling. Take the A1198, but now use its original Roman name, Ermine

Street, and it starts to come alive. I would suggest you go to Caxton and then head south, towards Royston. The road here lives up to its Roman engineers' dream: it runs dead straight, for miles and miles. As you pass through rural woodlands, the trees have been cleared back, just as they were in the late seventeenth century, giving a wide field of vision, discouraging lurking highwaymen. There are a few Georgian wayside inns, all set back the regulation distance from the road. It's a lightly travelled route, and one would not be surprised to see a coach and horses approach from the other direction. If ever there was a path to the past, Ermine Street has to be the one.

16

A Bridge Without Sides

The Old Lower Hodder
Packhorse Bridge, Lancashire

The lightly wooded landscape on either side of the River Hodder, near Clitheroe in Lancashire, has remained largely unchanged for many centuries; and I suspect my first view of the Old Lower Hodder Bridge would have been instantly recognizable to somebody who had used it in the decades after its construction in 1580. It is known as the Old Lower Hodder Bridge to distinguish it from the neighbouring road bridge and boundary between Yorkshire and Lancashire, the Lower Hodder Bridge, which was built in 1819. The old bridge also acquired the name Cromwell's Bridge when Oliver Cromwell used it to move troops north in the lead-up to the Battle of Preston, in August 1648: a major victory for Cromwell that would prove one of the decisive moments in the English Civil War.

The bridge was built by the prominent Roman Catholic Sir Richard Shireburn (or Sherburne), owner of the nearby country house of Stonyhurst Hall, whose park extended down towards the River Hodder.[1] There has been a house on the site of Stonyhurst Hall since the fourteenth century, but Sir Richard enlarged it considerably, adding a fine gatehouse in 1592. Cromwell is reputed to have slept on a table in the hall, in full armour, on the night before the Battle of Preston. It would seem he didn't altogether trust his Catholic hosts. Sir Richard's descendants continued to build additions to the Hall into the early eighteenth century, when it was inherited by the absentee Dukes of Norfolk, who never lived there. The house fell into disrepair and was given to the Jesuits in 1754. Today Stonyhurst is a well-known Roman Catholic public school.

From the side and from a distance, the Old Lower Hodder Bridge looks extraordinarily slight. This is doubtless because it has no walls – something that is very unusual on a stone structure. Indeed, when I first caught sight of it, I couldn't believe my eyes: it seemed as if somebody had deliberately damaged it; nothing so slender could possibly have survived for so long. Just over six feet wide, the bridge consists of three arches with a broad central span. The tall, very wide central arch gives the bridge a strangely modern appearance, despite its great age. A bridge of such a broad span has to be maintained in good condition, so I was concerned to read on the Historic England website that it was now suffering 'extensive significant problems', brought about by natural erosion.[2] In these times of increasing flood-risk a weakened bridge would be very vulnerable to an upland river in spate.

The Old Lower Hodder Bridge is one of the finest packhorse bridges in Britain.[3] In the previous chapter, I discussed the rise of turnpikes in the later seventeenth century along the existing network of formal roads – roads that were considered to be the king or queen's highway. As well as these formal roads, Britain was crisscrossed by networks of informal roads. These included drove roads: generally wide and straight, they were used to drive cattle and sheep to markets in towns and cities. Markets like London's Smithfield, for example, were largely kept supplied by specialist drovers, who under-stood livestock and knew how to drive them over long distances without the animals losing too much condition. In many parts of the country, especially in hilly areas where canals and, later, railways could not readily be built, local industry was kept supplied by strings of packhorses, who carried their loads in panniers slung on either side. The packhorses would often work in tethered teams, supervised by a man, or men, on foot. Packhorse roads played an important part in the development of the growing Lancashire textile industry. Bridges on the packhorse road network had to make allowances for the low-slung side panniers, which hung down a foot or two above the ground. This explains why the Old Lower Hodder Bridge has no side walls.

The drovers' inns, meanwhile, were far more basic than the smart hotel-like buildings along the turnpike network at Caxton and else-where. They were often located out in the open country, far from larger settlements. Their signs were often distinctive too. The old drovers' inn next to our small farm in the Fens was called 'The Gate Hangs High': its sign, which used to hang from a tree, was a small wooden gate.[4] The drovers' inns didn't have stables for replacement horses; rather, animals could be kept safely overnight in a hedged paddock attached to them. The drove and packhorse roads continued to thrive well into the nineteenth century, and the early railway age, because they were fare- and toll-free. They operated very effectively, filling important gaps in what today we would call the supply chain.

It might be supposed that in the early industrial era, transport and logistics were necessarily simple and unsophisticated. In actual fact, the opposite was the case: supply had to meet demand – and just like today, customers' requirements were never simple, nor straightforward. Awkward natural barriers, such as rivers, had to be dealt with by the people who were most directly affected and it wasn't as easy to tap into banks and government funds as in more modern times. Local trade and commerce could not be ignored, even by the rich and powerful. Ultimately it was these market forces that, back in the 1580s, inspired the building of this simple, elegant and functional bridge. It has earned an honourable, and I hope long, retirement.

17

Edinburgh New Town

A Vision and Its Realization

Edinburgh is one of my favourite places. Its layout, like that of most British cities, conceals a wealth of different phases and developments, some of which were more successful than others. Perhaps the best-known single episode in the development of any major British city was the planning, and subsequent construction, of James Craig's famous Edinburgh New Town.

The Scottish Enlightenment of the eighteenth and early nineteenth centuries was astonishingly influential. It had a profound effect on rationalist thinking not just in Britain and northern Europe, but across the Atlantic, thanks to the Scottish diaspora and the maintenance of close links with the American colonies. It helped give birth to the modern disciplines of political science and economics, and played a major role in the flowering of medicine, mathematics, natural and even early social science. Literature and poetry flourished too, as exemplified by James Boswell and Robert Burns. For all this, the capital city of this most civilized of modern democracies remained resolutely medieval, known to its inhabitants as 'Auld Reekie' ('Old Stinker'). Development was confined within the city walls, and many houses became multi-storeyed – and not always hygienic or stable. Trade, industry and commerce expanded with difficulty. So, in 1752, as part of a campaign to improve commerce, the central Convention of Royal Burghs drew up plans for Edinburgh's expansion. These were successful: so much so that by the end of the eighteenth century, the city was being widely dubbed 'the Athens of the North'. The building of the New Town played a major part in this astonishing transformation.[1] But it was not a straightforward process. As we still find today in the world of property development, the actual construction of the New Town was a difficult process, in which the ideals of planners and the need of developers to turn a profit for their investors often came into direct conflict.

In 1766, Edinburgh Town Council announced a competition to design a New Town to the north of the Old. The Old Town, the medieval and Reformation city, grew up on raised ground at the foot of a very steep hill, which was all that remained of an ancient, extinct volcano: like many prominent hills, it was probably first occupied in Iron Age times. By the twelfth century, this fortified hilltop had become a royal castle, which overlooked and provided protection for the settlement that was arranged around a central street, the Royal Mile, which ran east from the foot of Castle Hill, along a low natural ridge towards the Abbey, and later the Palace of Holyroodhouse. The ridge on which the Old Town grew up was bounded, and naturally defended, to the north and south by areas of marsh and open water.

In 1460 the marshy area to the north, known as the North or Nor' Loch, was deepened and damned on the orders of James I, to strengthen the town and castle's northern approaches.[2] But throughout the 1600s and early 1700s the Nor' Loch accumulated vast quantities of rubbish, which also blocked the city's northern expansion. In the 1760s, the draining and clearing of the Nor' Loch was finally undertaken and the City Council felt able to start the process of planning a New Town.

The New Town's designer, James Craig, was a highly gifted architect, who never rose to national prominence and died in very reduced circumstances. His New Town plan (of 1766) is undoubtedly his greatest gift to posterity, and he won the Council's contest against considerable competition from no fewer than five opposing plans. Such was his penury in later life that he even tried to sell the medal he had been given for winning the City Council's competition.

Craig's design was innovative, but not revolutionary; he plainly intended it to be effective and practical. As well as a clear vision of how the development should look from the outside, he had a specific idea of how its inhabitants should view the city and the landscapes beyond. So he specified building heights and defined which were to be the 'principal streets' and how they should be serviced by mews

and 'back streets'. Bearing in mind the need to see beyond the confines of the New Town, he clearly specified that the the two principal
streets defining the northern and southern edges of the New Town,
Queen Street and Princes Street, should not have buildings along
their outer sides. We can see the results today: from Princes Street,
you can enjoy the great views that Craig had in mind, the Castle and
Old Town across the wide green expanse of Princes Street Gardens –
which themselves are laid out on the drained, filled-in Nor' Loch. To
the north, from Queen Street, the panoramic views also remain,
across the city to the Firth of Forth.

As with the Old Town, Craig's precise positioning of streets in the
New Town reflected the underlying geology: a low axial ridge, along
which the central of the three principal streets, George Street, was
aligned. From George Street the linking cross streets run gently
downhill to join the other two principal streets to north and south,
Queen Street and Princes Street. It was a clear and simple plan, but
it would prove impossible to achieve in its entirety.

As I suggested earlier, there were problems. Developers didn't
always follow Craig's precise instructions and the City Council, worried about discouraging the inflow of finance and capital, were slow
to call them to account. A private mansion was built in the space
Craig had reserved for a principal church, St Andrew's. Some development was even allowed on the south side of Princes Street, which
partially blocked the wonderful views to the south. These things
infuriated local people and the Council was forced to buy back the
plot of land on to which it squeezed St Andrew's Church. In the
1780s, further efforts were made to force developers to keep to
Craig's plan, but even so, unregulated building continued. Things
came to a head in the 1790s, when piecemeal development of the
western end of the New Town clearly threatened to undermine
Craig's original vision of a well laid-out square. Belatedly, the Council came to their senses.

They commissioned the acclaimed neo-classical architect and
interior designer, Robert Adam, to design an entirely new square,
renamed Charlotte Square. Adam, the son of William Adam (1689–
1748), the leading Scottish classical architect of the first half of
the eighteenth century, completed his design in 1791. He died the

following year, just as building work was starting. His large, tree-lined garden square was completed in 1820, and this time the builders remained true to Adam's original design. I am a great fan of Robert Adam's work. He has a lightness of touch combined with an infallible eye for elegance, and Charlotte Square exemplifies both. It remains a triumph of classically inspired urban architecture and, for me, entirely justifies Edinburgh's title of the Athens of the North.

For many visitors, Edinburgh is still the city of the Old Town, the Festival and the military tattoo in the Castle, high on its spectacular rocky pinnacle. Craig's New Town is very different, distinct and set apart. Crossing Princes Street Gardens after the confines of the medieval city, you feel as though you are entering quite another world. Its straight streets and spacious pavements (Craig specified they must be at least ten feet wide) feel part of something else, something truly international. While the grey stonework may be local, the design of the buildings is essentially European. These elegant, yet essentially modest, buildings symbolize Scotland's membership of a much larger social and intellectual community. The New Town disdains insularity. It grew up outside the old defences and seems to reject the hints of paranoid confinement that made the Old Town so squalid in its latter years. It proclaims a far broader, more rational, and therefore more humane, world view. We are living through increasingly polarized times, when Britain seems uncertain about its future. Edinburgh New Town should remind us that differences can always be resolved, providing there is a will to do so. It remains a profound inspiration, and an elegant symbol of Enlightenment.

18

Perfection at Rousham

Kent in Oxfordshire

The county of Oxfordshire is blessed with some superb country houses set within beautifully landscaped parks. Everyone knows about the magnificence of Blenheim Palace, which was given by a grateful nation to the first Duke of Marlborough for his successful military campaigns of the early eighteenth century. It's surrounded by a wonderful park, largely designed by the famous 'Capability' Brown and it is visited daily by dozens of coaches packed with eager visitors. I have to confess, however, that I don't always feel at ease when surrounded by such opulence. I prefer something more restrained, that's closer to nature and reflects more than power and glory. My ideal English landscaped park and garden is to be found just six miles north of Blenheim, at Rousham House, near Bicester.

Rousham lies towards the eastern edge of the Cotswold Hills, some ten miles east of Chipping Norton. It's a busy part of the country, but the rural landscape around here isn't crowded; most of the traffic seems to be drawn on to the M40, which passes quite close by. It's a classic Cotswold scene of small limestone villages, drystone walls, woodland and grazing, with arable fields on the open, flatter ground. The house and gardens at Rousham are well positioned on the edge of the valley of the River Cherwell, a major tributary of the Thames, which it joins at Oxford. At Rousham, the Cherwell is a small river and forms an important part of the garden's framework. What is it about English landscaped parks and gardens that makes them so special? My own view is that they are firmly rooted within their landscapes, settings from which their designers have always drawn inspiration. As a result, there is a constraint and humility about their work that I find very appealing, and which is sometimes absent from the grand designs of continental Europe.

The English tradition began in the late Middle Ages, thrived in Tudor and Stuart times, and reached what many would see as its first peak in the early eighteenth century, with further achievements in Victorian times, with the Arts and Crafts movement of the early twentieth century and, indeed, today with, for example, the superb garden of the late Christopher Lloyd, at Great Dixter, in Sussex. Rousham is an essential part of the development of what I've just described as the first peak of English park and garden design, in the eighteenth century. Here you can see most of the ideas that inspired the later, and much larger, designs of 'Capability' Brown and his successor, Humphrey Repton.

Later medieval and post-medieval garden designs reflected the prevailing social order and their tastes. They were laid out for people with large country houses and budgets big enough, almost literally, to move mountains – and certainly to divert or dam a few streams. Initially gardens in the fifteenth, sixteenth and early seventeenth centuries were laid out to glorify a monarch, should he or she choose to visit, or to proclaim the importance of an aristocrat and to embellish the place where he, or she, resided. Many of these grand, formal gardens were open to genteel visitors rather than the general public.[1] Tips paid by such visitors to senior servants provided a welcome boost to their income. Although I described the visitors as 'genteel' it is probably fair to say that most country houses were less exclusive than they are today, and local people would have had access to the parks, if not to the private gardens.

As time passed, the strict formality of early post-medieval gardens began to break down. Rigidly geometric parterres – a form of garden characterized by gravel paths and closely cut box hedges, which developed in continental Europe – began to be accompanied by less formal areas of lawn, where the aristocratic family could relax, away from the public gaze. By the later seventeenth century the process was gathering pace, as new money, either acquired in the growing world of industry or from trade in the expanding overseas empire, started to make itself felt. But these were the years of the Enlightenment, when it became less acceptable to draw attention to one's own importance in too obvious a fashion. Status for its own sake mattered less, and other aspects of a man or woman's personality gained in prominence. The emphasis shifted towards intellectual pursuits, learning and creativity: music, painting, architecture and horticulture. So we start to witness the introduction of an entirely new style of garden design, which the historian Tom Williamson has memorably described as 'Polite Landscapes'.[2] Essentially, polite landscapes are all about the display of a landowner's good taste, and his erudition – especially in the classical philosophy, myths and legends of ancient Greece and Rome. They were intended to arouse quite specific emotions, for example of awe, excitement, humility or triumph, depending on the statues and scenery being deployed.

Today eighteenth-century landscape gardens, with their sculptures, scenes, flowerbeds and vistas, have an enduring appeal, irrespective of the fact that the gardens' original significance and purpose are often lost on us. We rightly take them at face value, as gardens. And they can take this scrutiny well, because they were laid out with great sympathy for their surrounding environment. The great eighteenth-century British garden designers sought to modify the landscape and display its capabilities, rather than to transform it through brute force: hence the name of the best-known of these men, Lancelot 'Capability' Brown, who arrived on the scene at Stowe, in Buckinghamshire, in the 1740s. Stowe isn't so much a garden as a designed landscape of enormous historical importance, as it illustrates all the significant developments of the eighteenth and early nineteenth centuries. Brown worked there at the start of his career, under the excellent supervision of William Kent, the designer, as we will shortly see, of Rousham gardens. Brown became the leading designer during the glory days of the English landscape movement. He created many parks and gardens, and some 170 survive to this day.

The garden at Rousham was designed by Brown's predecessor there, William Kent. Kent's designs were more restrained and more classically inspired than Brown's and personally I prefer them. Unlike Brown, whose background was in parks and gardens alone, Kent was also highly regarded as an architect and interior designer – and you can see this in his gardens. The grounds at Rousham House are widely regarded as his masterpiece.[3] We are fortunate that the family that built and owned Rousham House, the Dormers, and, after 1719, the Dormer-Cottrells, immediately recognized that in the gardens they had acquired a work of genius. They have curated them with affection and great skill ever since.[4]

The house was built around 1635. Positioned on the slopes of the Cherwell valley, with views across the floodplain, the grounds are bounded by the river itself. The gardens were first laid out by Charles Bridgeman, in the more naturalistic style that was then becoming increasingly popular. Designers didn't want to remove their newly created gardens from the surrounding landscape, so 'invisible', so-called 'ha-ha' boundaries that were livestock-proof and formed by sunken walls, were introduced to Britain by Bridgeman. These features were elegant when seen from outside the garden, but were almost invisible when viewed from inside, or from the house. Bridgeman used the ha-ha very effectively at Rousham to separate the gardens from the park, which was grazed by cattle.

Bridgeman finished his work in 1737. The following year, James Dormer-Cottrell commissioned William Kent to develop and enhance Bridgeman's earlier work. This he did with extraordinary confidence that recognized the good points of the simplicity of the earlier scheme and built on them, creating many new and imaginative features, such as streams, temples, springs and grottoes. In just four years, he had made a new garden that acknowledged Bridgeman, but was one hundred per cent William Kent. Kent's design uses surrounding views and the adjacent river with great skill. Each time you glimpse the river, it's as though you're seeing it afresh. On the uphill side of the garden is the park, with some fine old oaks and a herd of English Longhorn cattle, which are kept out of the garden by the earlier ha-ha.

William Kent designed his garden to draw visitors along a particular route: we know what it was because his plans survive with his other papers. When I first visited Rousham, however, I didn't know about the route and missed some of its intended effects as a consequence. Having said that, ignorance can sometimes be bliss. So we wandered at random: turning into interesting dark corners, and stumbling across hidden delights. One of these was the Rill, a carefully channelled stream running through dark woodland, which opened up to us unexpectedly. By it ran a path, which took us through deep woods and then slowly widened, to reveal a temple and a wonderful vista across the floodplain. I think it was that first unplanned visit that captivated us, yet that was not how Kent wanted us to experience his garden. When we came back (something we have done repeatedly) we followed his route, and got even more from each visit. And that, surely, is the measure of Kent's genius: his gardens work in whatever way you choose to experience them.

No garden of this kind would be complete without some set-piece scenes. The Bowling Green, the large lawn that extends from the formal front of the house across to the brink of the valley slope, was originally conceived by Bridgeman and Kent had the good sense to leave it unaltered. It uses the handsome house very subtly. It is not a showy building and its Jacobean style lacks the lightness and elegance of William Kent's time. In essence, the lawn and the two huge yew hedges that bound it on either side, draw the eye away from the house and out across the Cherwell valley. Here the nearby landscape has been 'borrowed' – visually taken into the park in order to expand and enhance set-piece views – by the inclusion of a decorative Mock-Gothic cottage and a Gothic-arched road bridge, which still carries the public highway. Many of the great parks and gardens of the age 'borrow' landscapes in this way. At Stowe, these long-distance views radiate several miles out into the surrounding countryside; at Rousham the borrowed views are more restrained, more intimate.

If you return to the house or look for a small path through the huge yew hedge towards the stable block, you will find yourself in a different world. The path opens up to reveal superb borders in an early walled garden, and the formal bedding around the elegant circular stone dovecote is a delight. Much of this is more modern work, but it is highly competent, with some very old pleached fruit trees. The planting of the borders in the walled garden is a delight and the box-edged beds around the dovecote hark back to pre-Kent times. The nearby parish church is very effectively 'borrowed' as a background setting for some of these plantings. Despite this careful planning, nothing feels contrived; all seems organic and in harmony. These smaller gardens are very different from Kent's thought-provoking and imagination-expanding creation across the great lawn. And it's the contrast between the two that is so appealing: intellectual formality, versus domestic relaxation. If I was looking for a single word to describe Rousham's subtle appeal it would be this: it is unashamedly *English*.

19

The Causey Arch, Co. Durham

A Long-forgotten Record Breaker

Over-simplification: it's an essential, inevitable part of the creation of history, which is always structured around stories. So it is with the Industrial Revolution, which, we learn at school, began with the canals, followed by the first steam engines, then the Stockton to Darlington Railway – and before we can say 'James Watt' we have arrived at Euston Station and the first mainline trains to Birmingham. It's a cracking tale. The actual truth was rather more complex and even more colourful, as we will discover when we examine the remarkable history of an unusual bridge, outside the village of Tanfield, near Stanley, in County Durham.[1]

The conventional wisdom on the development of Britain's canal network tends to pass over those coal-producing areas of the country where the terrain was too steep to allow canals to be built and operated cost-effectively. One of these areas, the Durham coalfield in the hills around Newcastle upon Tyne, was highly productive and had good overland access to the sea. To take advantage of this route, mine-operators had developed their own systems of horse-drawn railways, called waggonways or tramways, which were in operation from the start of the seventeenth century, and probably began much earlier. The tramways transported coal from the uplands down to the ports at Newcastle and Gateshead, where it was loaded onto vessels and shipped south to London, where 'coals from Newcastle' had been keeping the population warm from medieval times.

By the eighteenth century, as industry gathered pace and the population of London rose steeply, the demand for fuel from the Durham coalfield was increasing and conventional road waggons were unable to cope with the higher output. So the mine-owners decided to build a rail waggonway, of a type long used within mines and quarries. The first recorded waggonway was built by Lord Middleton in 1600, to move coal from his mines at Strelley to the markets in central Nottingham, a distance of four or five miles. Over a century later, an ambitious new tramway in County Durham, financed by a consortium of mine-owners known as the Grand Allies, was to be much longer. But there was one major obstacle to their plans: the plummeting slopes of the steep-sided gorge of the Causey Burn, a tributary of the River Tyne.

I can remember driving through the Durham coalfield and the country around Newcastle upon Tyne as a lad, back in the 1950s. Strangely, my memories are in black-and-white – and perhaps they're false, ultimately derived not from my own experience but from monochrome news footage. In my mind is a landscape of slag heaps and terraced houses, mines and railways, with level crossings constantly closed while trains of wooden-sided coal waggons, hauled by black, grimy tank engines, slowly clank by. I remember a man standing on the small balcony of the guard's van at the rear of one train. Seeing my frantic waving out of the car's rear window, he raised his enamel mug in our direction.

This landscape has changed radically in the intervening decades. Returning a few years ago, I was completely taken aback: the grime and dirt, even the railways, had vanished, and with them the all-pervasive smell of smoke on the air, from houses, factories, workshops and steam trains.

Then, the open country surrounding the mines had been quieter too; presumably insects, mammals and birds struggled to thrive in so polluted an atmosphere. But now, as I walked along the path to the Causey Burn, it was like a stroll through a nature reserve: I could hear the rattle of a woodpecker in the trees, while high overhead two buzzards lazily circled on outstretched wings. As I looked around me, I became increasingly conscious of time's extraordinary ability to repair the damage we have inflicted on our world. Those soaring buzzards up there were at the top of the food chain, their existence entirely based on a vast supporting pyramid of bacteria, fungi, lichens, plants, insects and smaller birds and mammals. If they had been reintroduced into the area in the 1950s they would have failed.

But now they were thriving. I was beginning to feel optimistic about the future until I spotted the vapour trails of two huge jets, high above.

The Causey Burn, a stream rather than a river, hardly sounds like a major obstacle. But it flows through a steep gorge that had somehow to be crossed by the waggonway, as bypassing it would be impractical. Small upland streams can become powerful engines of erosion when in spate, and this one had cut an impressive gorge, with vertical, crumbling, rocky sides. There was no alternative than to bridge it, and construction began in 1725.

The man who had been handed the task of traversing the burn was stonemason Frank Woods. Basing his design on simple Roman principles, Woods built a stone bridge whose single-span arch was, at 102 feet, then the longest in Britain. It never failed, although Woods – a nervous man – lived in permanent fear of his structure's collapse. Eventually this dread caused him to commit suicide.

The loads the bridge had to bear were formidable. The coal was transported along a railway, the Tanfield waggonway, in horse-drawn, four-wheeled dumper trucks, each of which could carry 2.65 tons of coal. The scale of the trade in 1725 – a good century, remember, before the appearance of the first recognizably modern steam railways; an era in which ladies and gentlemen wore wigs and gowns and the spa town of Bath was the height of fashion – was enormous. Woods' bridge was crossed by some 930 waggons each day – that's one waggon every twenty seconds. The railway's wooden rails were so heavily worn down that they required replacing every year.

The Arch is quite close to the highest point on the waggonway, on the landward side. Here, laden timber trucks, known as chaldrons, were towed, about 160 feet apart, up the slope. Before starting the downhill run to the sea, the horses would be detached and re-harnessed behind the chaldrons, to act as a brake for the long slope down to Newcastle. Once the load had been discharged and packed (by hand) into the waiting colliers, one horse would pull four empty chaldrons back to the mine on the tramway's second, parallel track, known as the 'bye way'; loaded waggons followed the 'main way'.

The Causey Bridge was heavily used for about eight years after its opening in 1726. Then, from 1733 to 1738, many of the smaller mines it served were closed down. Iron rails replaced the wooden track in 1739, but the following year a disastrous fire closed its main source of traffic, the highly productive Tanfield Colliery. After 1770, the line and the bridge went out of use. It was in urgent need of stabilization and reinforcement when, some two centuries later, in 1981, the County Council started their extensive restoration work.

Many bridges are in spectacular locations that dominate the surrounding landscape. Not the Causey Arch, which seems to lurk rather modestly around a bend in the gorge. After looking at the bridge from the top of the incline, I left the path and made my way down to the bottom of the ravine, alongside the Causey Burn. A steep buttress of rock and thick vegetation obscured my view of the bridge, but when I rounded the rock I was taken aback. The bridge, far higher than I had imagined when seeing it from above, spanned the burn with a lightness and elegance that put me in mind of the Old Lower Hodder Bridge (see Chapter 16). Built to be functional, it nevertheless had an effortless grace. With over nine hundred waggons passing over it daily, not to mention the many horses and their drivers, I suddenly understood Frank Woods' fears: any fault in the stonework, at such a great height, would have been catastrophic. The Causey Arch remains a soaring memorial both to the talented engineer who conceived it, but also to the many, now anonymous, men who built, maintained and used it. It certainly deserves to be much better known.

The Causey Arch viaduct, Co. Durham.

20

Birkenhead Park

The First Public 'People's Garden'

Brought up in the country and accustomed to a less hurried pace of life, I have never found it particularly easy to live in cities. Even when young and drawn to London nightlife, I would return home on weekends to work in my father's large garden, which I had helped to design, and to take long walks through the rolling chalk hills of north Hertfordshire. But then I emigrated to Toronto, Canada, and suddenly I had to learn how to live an urban lifestyle. Toronto might be one of the best cities in which to do so. It's blessed with some fine parks: the formality of Queen's Park, close by the Royal Ontario Museum, where I worked, and a huge Victorian cemetery with vast Canadian maples, directly behind our road. After nine winters in Canada, I was on nodding terms with every squirrel and racoon in those parks.

Unsurprisingly, my stay in Toronto fuelled my interest in parks. Urban green spaces are the things that make town life tolerable for most of us. Nobody knew this better than Joseph Paxton, the man who designed the first publicly funded urban park, anywhere in the world, in the Merseyside town of Birkenhead. Countless others since have also recognized its importance, for it is, still, remarkably intact.[1]

In the past, in common with many other urban parks, Birkenhead Park has had to endure spells of neglect; more recently, however, it has been fastidiously restored – and is now resplendent. The history of Georgian and early Victorian Birkenhead is an extraordinary story of local enterprise and enlightened foresight. At the start of the nineteenth century, Birkenhead was still a small village, in the shadow of its vast neighbour across the Mersey, the port of Liverpool. Then, in the 1820s, industry, especially shipbuilding and engineering, slowly became established there. A local businessman and entrepreneur had commissioned the highly regarded Edinburgh architect James Gillespie Graham to lay out a new development to rival James Craig's much-admired Edinburgh New Town, which, indeed, Gillespie Graham had helped to fashion, having laid out a part of it, the district known as Drumsheugh, in the early 1820s. One surviving incarnation of Gillespie Graham's development plan for Birkenhead was the superb and surprisingly uncelebrated Hamilton Square, which was built between 1825 and 1844. It is not generally appreciated, but it has the greatest number of Grade I listed buildings anywhere outside London.[2] Its four-storey, stone-built Georgian terraces exude grace, confidence and authority, and were doubtless built to attract up-market clients to the town.

Work on Birkenhead Park, an integral part of Gillespie Graham's overall plan, began with the setting-up of a local Improvement Commission, which raised public money to buy 226 acres of marshy land to the west of Hamilton Square. More money was raised by selling off plots of land for smart new houses around the edges of the proposed park. Eventually, in 1847, sufficient cash had been raised to commission designs.

Any account of Birkenhead Park has to start with its remarkable architect-designer. Joseph Paxton – Sir Joseph, as he became – was best known for his greenhouses and glass buildings, including the Crystal Palace, the massive structure that housed the Great Exhibition of 1851. (Paxton was also a plantsman, famous for breeding the Cavendish banana, which we eat in vast quantities today. He named the banana after his employers at Chatsworth House, where he was Head Gardener; this, in fact, was his day job, while designing the likes of Birkenhead Park and the Crystal Palace.)

While you might have expected Paxton's designs for Birkenhead Park to have reflected his ongoing work at Chatsworth, they didn't. In fact, it would be hard to find two more contrasting projects. Chatsworth is internationally renowned for the magnificence and grandeur of its grounds and gardens, with miles of tightly clipped hedges, acres of trimmed grass, glasshouses and, of course, that famous fountain.

Glasshouses and fountains, however, are entirely absent from Birkenhead Park. Indeed, on my first visit to Birkenhead I was struck, after walking through the classical Grand Entrance, by the sense of – entirely manmade – rural tranquillity that Paxton managed to achieve. This is Paxton in demotic mode, making no attempt to over-awe with tree-lined avenues and towering fountains. Rather, he designed the park as a – delightful – piece of everyday life: an ideal domestic garden, to be enjoyed daily by as many people as possible. (Some 10,000 locals turned up to the park's opening in April 1847.) Visitors find their way around the park following paths and walks that meander through it, revealing its views – the boating lake, for instance, and the fine columned Roman Boathouse – in an unexpected, completely 'natural' fashion. A demonstration of Paxton's remarkable versatility, it could not be a greater contrast with Chatsworth.

Even the scale of the park's buildings is entirely appropriate. The two-storey Grand Entrance, a triumphal arch in miniature, feels somehow domestic; meanwhile, the red-painted, roofed and recently restored wooden Swiss Bridge at the edge of one of the two lakes has a lightness about it, almost disappearing into its surroundings.

As a practical gardener myself, I find Paxton's landscaping sleight-of-hand extraordinary. In fact, painstaking engineering work was needed to create the park. The marshy land was naturally damp and poorly drained, so many pipes had to be laid to remove floodwater. These subsequently silted up and had to be restored in the recent work. The park's two large lakes would have helped the drainage, but would never have coped on their own; moreover, their water levels were not intended to fluctuate, or the naturalistic edge planting would simply not have worked. So their levels needed careful control – and that took skill. There are no visible sluice gates or balancing ponds; indeed, the park manages to conceal its high-quality engineering with wonderful modesty.

Paxton's Birkenhead Park triggered a spate of urban park construction. The nearby, slightly later Sefton Park in Liverpool features a picturesque lake and a very fine Paxton-style, three-tiered glass Palm House. One admirer of Paxton's achievement at Birkenhead was the American landscape architect Frederick Law Olmsted, who loved everything about the park, as well as the public use for which it was designed. It was, he said, 'perfection', an incomparable 'People's Garden'. There was 'nothing' like it in America – that is, until Olmsted himself designed New York's Central Park, which, at three times the size, is the most celebrated of the parks inspired by Paxton's Birkenhead.

21

Risehill Navvy Camp

The Smell of Death, Above the Clouds

The construction of Britain's rail network, which mostly took place in the 1840s, was a Herculean task; many companies failed, and many banks along with them. The process produced some notable architects and engineers who bequeathed us a wonderful legacy of bridges, viaducts, stations, hotels and housing. However, the people who actually did the work have largely been ignored, written-off as mere 'navvies'. Navvy, of course, is shorthand for navigator, the name given to the men who dug the canals of the previous stage of our industrial evolution. The navvies preferred to think of themselves as 'railway labourers' or 'excavators'. They had a reputation for hard living and excessive drinking, and well-meaning – if very patronizing – efforts were made to have them adopt more temperate ways of life.[1] If they really were the drunken oafs of popular history, though, you wonder how on earth they managed to construct most of the tunnels and bridges of our railway network in just ten years? So what was life really like on a big construction project?

By modern standards, working conditions were often appalling and accidents were very frequent. We know that death rates on these projects were higher than we would accept today. Yet we know, too, that many engineers, constructors and major contractors, especially Samuel Morton Peto, who was one of the principal construction contractors, employing some 9,000 men in 1846 at the height of the railway boom, cared deeply about the well-being of their employees, while the living quarters provided by the larger companies for their workers improved dramatically after the 1840s.[2] Such incentives perhaps helped create dynasties of navvies: many navvy fathers had sons who became navvies, too.

The great 1840s expansion of railways mostly took place across populous parts of Britain, where something as ephemeral as a navvy camp would be unlikely to survive: temporary structures, such as huts and shelters, leave very slight and shallow traces in the ground that are easily obliterated by even light cultivation or development. So, to try to unearth remnants of a navvy camp, I knew I had to find a site in an area where arable farming was impossible. During the first decade of the present century I was quite heavily involved with the Channel 4 series *Time Team*, and suggested that we look for an upland railway built some three decades after the main expansion of the 1840s, but using essentially the same techniques: the work of digging and tunnelling still being done by navvies and not yet by mechanical excavators. Happily, one such line remains intact: it is known today as the Settle–Carlisle Railway.

The story behind the railway's construction doesn't throw a particularly kind light on mid-Victorian corporate governance. In essence, it was all about a particular company's vanity. The Midland Railway disliked having to use other railway firms' tracks to take their trains to Scotland. So they decided to build their own route across the high Yorkshire Dales and North Pennines. It was a decision that was to cost the lives of more than a hundred navvies, mainly through disease and accidents, but what they achieved in a mere seven years – construction began in 1869 and was completed by 1876 – was phenomenal: seventy-three miles long, the railway crosses some of the harshest upland terrain in England, which is why it involved the construction of fourteen tunnels and twenty-two viaducts.

The engineering jewel in the line's crown is the magnificent Ribblehead Viaduct, whose twenty-four arches so elegantly span Ribblesdale, high in the Yorkshire Dales. To one side of the viaduct is the site of the Batty Green navvy camp, on Blea Moor Common. Eighty people are known to have died there during a smallpox epidemic. I first visited Ribblehead on a summer's day in 2005, after several days without any rain. An area of coarse grass on a low rise of dry ground near some craggy outcrops of rock was looking especially parched and patches of deep peat stood out starkly. I remember standing on some of the low humps and bumps that were probably the collapsed remains of rough, stone-built fireplaces and chimneys that heated the navvies' huts. In the sandy soil at my feet I could see a few tiny scraps of glazed pottery. I looked towards the viaduct across what seemed like acres of disturbed ground. The huge site, which had held 2,000 navvies at its height, was essentially intact. It was also a protected Scheduled Ancient Monument and I knew it would be wrong to disturb it. So where else could I turn?

A few years later, in the summer of 2008, along with a film crew of fifty, I found myself at over 1,200 feet above sea level, atop Risehill, directly above the Risehill Tunnel, a few yards down the line from the Ribblehead Viaduct. The weather that July was unsettled, and we frequently looked down on the rainclouds above Ribblesdale. Risehill itself was bleak and treeless: there was nothing to protect us from whatever the weather had to throw at us. Our supplies came by four-wheel drive, or helicopter.[3]

As a practising field archaeologist, I am used to working in areas that have been transformed by the hand of man, and I rarely get an opportunity to experience places that at first seem to have returned completely to nature. Risehill was just such a place – or so I thought when I recce'd the site a month before the dig began. Almost everything about the place looked untouched by human activity: all around were tussocks of tough marshy grass and huge puddles, pools and wet places where sphagnum (bog) moss was accumulating in thick spongy bands. It all looked *so* natural, but I soon realized that it wasn't.

Near the apex of the hill were a couple of small circular brick walls, like very large well-heads. These were the tops of the vertical air shafts that had been sunk both to ventilate the railway tunnel and as a means of hauling quarried lumps of rock up to the surface. It took me quite a long time to realize that in the early 1870s much of Risehill's upper slopes had, in fact, been covered with huge blocks of rock that had been blasted out of the tunnel, deep inside the hill. Far below the ground they were drilled, packed with explosive and blasted away from the tunnel face. Then, still weighing a ton or more apiece, they were winched and manhandled to the surface, up the shafts. Lying on the ground all around me, I could readily spot fragments of the two- or three-inch diameter boreholes into which the explosive had been packed. It must have been very hard and

extremely dangerous work. The rocks were then loaded into waggons and hauled away from the shafts, along temporary railway
lines, to where they were dumped, forming several dramatic ridge-
backed spoil heaps.

Through the late nineteenth and twentieth centuries, the rock
heaps had been gradually covered by lush growths of damp-loving
mosses, lichens, grasses and sedges. Some distance away were a few
low mounds which, rather like the rock heaps quarried from the tunnel, seemed to be protruding from the thick mat of grass and sedge.
On closer inspection we found fragments of rough masonry and
some coarse brickwork, and after a couple of days' careful excavation we had revealed the base of a fireplace. We knew a certain
amount about the later navvy huts from written accounts, which
suggested that the contractors for this part of the Settle–Carlisle line
had bought a collection of prefabricated barrack huts left over from
the Crimean War of 1853–6. The stone fireplaces, together with a
small brick-built extension, would have been added to the dormitory
huts to provide a living room and warmth.[4] Hardly palatial – but
still, a good deal better than nothing, at such an altitude.

On the second and third days of the dig it rained: a wall of rain
like I had never experienced before. It was impossible to take photographs or do any excavation without some sort of shelter. As
professional excavators we were used to harsh conditions, but even
we began to feel dispirited by the unremitting weather. Strangely,
though, I found it bringing me closer to the place and to the men who
had worked here, day in, day out, for at least three years. The bridges
and tunnels were one thing, but barely a mile of the line does not run
through a cutting or over an embankment. We take them for granted
today, but all of that earth had to be dug by hand, then carted, and
often in the most appalling weather.

We know from the national census of 1871 that four accommodation huts were located near the shafts down to the tunnel; by the end
of the three-day dig we had managed to locate them all. We also discovered the cinder paths that linked the huts together and allowed
access to the shaft and the winding gear. In fact, once we had
removed the covering of grass and sodden peaty topsoil to expose the
original paths, we found they still shed rainwater superbly, making it

easier for us to move around the excavation site. Given the navvies' reputation for drinking, we expected to come across broken bottles everywhere, but in fact they were rare. We found broken plates and tea cups far more frequently.

The census also reported that one navvy had committed suicide – the cause of death was not given – while sitting alone on the latrine, which we were keen to locate as part of the broader picture of the camp. After finding evidence for drains and soakaways down the slope from the main huts, which is where we would have expected the latrines to be placed, three of us began excavating; soon, we came across what was probably a holding cess-pit at the head of a steep and long downhill slope. Normally cess-pits on archaeological sites are odourless, because fungi and bacteria have long since digested any surviving faecal matter. But this was quite a recent site; moreover, it had remained more or less permanently waterlogged since the 1870s. The acidity of the rain would also have been very unfriendly to bacteria and fungi – with the result that the cess-pit smelled strongly. The smell was extraordinarily evocative: a familiar, obviously human smell, but strangely I didn't find it even slightly unpleasant. Quite the opposite, in fact. Maybe it was that lingering odour, but as I scraped away at the ground, I found my mind returning, again and again, to that poor man who had taken his own life. What a lonely way to go.

A footnote. If you ever find yourself at the huge viaduct at Ribblehead, I suggest you also make your way to the small nearby church at Chapel-le-Dale. The graveyard contains the bodies of navvies and their family members who died of disease in the camps, and there is a small inscribed monument to them in the church. The graves are unmarked, but if you examine the churchyard wall from the road, you will see a change in the brick- and stonework: in the 1870s, the cemetery had to be extended to make space for the sudden influx of new arrivals from Ribblehead. But for me their lasting memorial is not in this humble churchyard, but out there in the high Dales, where the towering arches of the Ribblehead Viaduct still stride across a bleak and rain-soaked valley.

22

Lordship Rec, Tottenham

The Countryside Contained

The British landscape had a very hard time in the mid-twentieth century.[1] Before the introduction of Town and Country Planning Acts in the 1930s and 1940s, unchecked development blighted huge tracts of green countryside around our towns and cities.[2] Pre-war suburban architecture was hardly innovative and very, very repetitive. Nevertheless, it did work. The streets were not as narrow as their Victorian predecessors, and many were planted with trees. The houses themselves were generally comfortable, well laid-out and many had gardens; large numbers of people were very happy in them – and still are. By the 1950s and 1960s, housing developments that were just a few decades old had become leafy middle-class suburbs.

During the decades of post-war rebuilding, planning controls became stricter, and one might have expected the actual architecture of houses and public buildings to have improved. Sadly, for the most part, this didn't happen. The building supply industry had become rundown during hostilities, when the focus was very much on munitions and armaments. Brickworks and kilns require constant maintenance, and quarries can soon become flooded, with the result that after the war there was a serious shortage of bricks, roof tiles and other conventional building materials. This led to the introduction of new components for buildings, which often involved prefabrication. These new materials suited the construction of high-rise blocks, which were seen as a way of helping to reduce the chronic housing shortage. It would, however, be a mistake to view the buildings and urban developments of the post-war years in historical isolation. With the possible exception of New Towns, such as Harlow, Stevenage and Milton Keynes, they were rarely 'planner's dreams' that appeared out of nowhere. Many took place as a consequence of earlier, pre-war developments, especially in those parts of the country that had been areas of rapid growth in the late nineteenth and early twentieth centuries. Tottenham, in north London, was just such an area.

On 22 July 1872 the railway station of Bruce Grove was opened in what is now the centre of Tottenham. It must have seemed like a strange arrival in what was still a semi-rural area. But just over a decade later, the Cheap Trains Act of 1883 made railway companies charge reasonable fares to all their passengers, which encouraged the growth of traffic to suburban stations including the then-developing outer suburbs of north-east London: Tottenham, Leyton and Walthamstow. At first, houses were built on good quality land near to railway stations and tram stops. But as time passed and the population grew, most of the accessible land was quickly built over. The remaining land was more difficult to negotiate. Many of the small streams and rivers in the countryside just north of mid-Victorian London drained into the River Lea (a major tributary of the Thames in Hertfordshire and east London). These streams flowed through wide, open valleys, which flooded at the smallest opportunity.

In more central parts of London, rivers such as the Fleet had long been buried in tunnels or conduits, a process that had begun in earnest in the late sixteenth century. But in the new outer suburbs, developers lacked the funds to carry out what was seen as inessential work. So they avoided the expense simply by not building on low-lying or flood-prone land. In hindsight, we can see how smart this was, allowing streams the space to accumulate and slowly dissipate surplus water. It's a practice that, today, developers and planning authorities often seem to forget or ignore, routinely siting housing estates in river floodplains.[3]

One of the few surviving farms in the maze of newly developed streets around Tottenham was the aptly named Broadwater Farm, whose lands were on either side of the flood-prone Moselle Brook (or River Moselle), a tributary of the River Lea to the east. The name, incidentally, has nothing to do with Moselle in France, and probably derives from 'Mosse-Hill', in Hornsey, the source of one of its streams. A remarkable late-nineteenth-century photo – taken, I have always assumed, to advertise a local milkman – shows three of his milk floats, and a solitary dairy cow, in standing water in Lordship Lane, which ran – and still runs – past what is now the well-known housing estate of Broadwater Farm. Judging by the casual attitude of the people and animals in the photo, this wasn't an exceptional flooding event, and it shows that part of what is now a major urban thoroughfare was once semi-permanently under water. The picture was taken in summer or autumn, to judge by the leaves on what look like lime trees beside the house – and for what it is worth, limes thrive best in wet, moisture-retentive soils.

Much of the land belonging to Broadwater Farm escaped the development that was triggered by the railway's arrival in the 1870s; in 1932, it was bought by the local authority, who converted it into the Lordship Recreation Ground, which opened to a rapidly growing population in 1936.[4] It's still there today: a gently sloping, open greenfield park with the River Moselle running through it, in a largely artificial cut. There are trees, some of them old pollarded willows, and small areas are given over to conservation. This is a park that is used: children and youngsters play football; a few drones fly about; dogs are walked and cycles are everywhere. The nearly fifty acres of Lordship Rec are an invaluable amenity and there is much local involvement in their management.

Shortly after its official opening, Lordship Rec became the beneficiary of a very unusual civic gift. In 1938, the then Minister of Transport, Mr Leslie Burgin, opened the Model Traffic Area, a scaled-down layout of streets where children could be taught the rules of the road – and how to stay alive in a busy city.[5] In its heyday, children could hire pedal-cars and bicycles on site; now they must bring their own. Like much of the Rec, the Model Traffic Area has been lovingly maintained by a large group of active volunteers, the Friends of Lordship Recreation Ground.[6] National Lottery and local council grants have been used to excellent effect. Instead of being straight and canalized, the course of the Moselle Brook is now gently meandering: its banks are less abrupt, fringed with rushes, and more natural – a striking contrast with the towering buildings of the Broadwater Farm Estate, immediately to the east.

Built in the late 1960s, the estate was based on the then-fashionable modernist ideals of the architect Le Corbusier, which centred on order and symmetry rather than the diversity of human society. It was also built across the floodplain of the Moselle and, consequently, the at-risk ground-floor level was entirely given over to car parking.[7] The residents of the estate essentially lived at the first floor, or 'deck level', mostly around a series of walkways that linked apartment blocks, shops, amenities and the high-rise buildings that still dominate the skyline.

The unusual layout and arrangement of the estate caused social problems, and these were made worse by official neglect. Early in the 1970s the estate acquired a reputation for crime, together with poor living conditions. Power failures happened frequently; rats and leaks and damp were commonplace. Local initiatives had started to address some of these problems, but, as so often happens, just as things had started to improve, disaster struck in the form of the infamous riots of 1985.

Subsequently, the estate was the subject of a major (£33 million) regeneration scheme. More importantly, its governance – formerly hugely unrepresentative of the estate's residential population – was democratized, to include representatives of the many diverse communities living there. Emphasis was placed on creating employment, and the drift of young people away from the estate was halted. Collectively, these reforms led to a steep decline in crime: today, indeed, Broadwater Farm has one of the lowest crime rates of any urban area. In 2005 the Metropolitan Police disbanded the Broadwater Farm Unit, the special force that had been assembled to keep peace on the estate. It was a simple action that showed confidence in the future and acknowledged what had been achieved so far. As I stood by the Moselle Brook in Lordship Rec and stared up at the towers of Broadwater Farm Estate, now half-concealed behind some gently nodding willows, I was reminded of what Joseph Paxton had demonstrated so clearly at Birkenhead a century earlier: urban parks and the houses that surround them can never be separated.

Lordship Rec, Tottenham.

23

Queensgate Shopping Centre, Peterborough

Opening Up City Centres

When I first visited Peterborough Cathedral in the 1960s, I remember thinking how dowdy it made the surrounding town appear. All the buildings were stained a greyish black by the coal smoke from the nearby railway. Traffic made its slow, noisy way through the Cathedral Square. Everything looked tired and run down, dispirited. Then, slowly at first, things began to improve. Buildings were cleaned. A huge ring-road and bypass were constructed, with carefully conceived views of the Cathedral. The 'Peterborough Effect' – a very clever advertising campaign to draw new business to the city, fronted by the popular actor Roy Kinnear – was a big success. And then Queensgate arrived, a building for which the city had been waiting: a civic structure that complements and enhances the Cathedral and which breathed new life into what is now a highly successful city centre.

Queensgate is magnificent, but, unlike the Cathedral, there are very few exterior views, and if you do manage to catch one it is soon forgotten. It is entirely based around its interior: something that rarely applied to earlier buildings. It is surrounded by a higgledy-piggledy accumulation of typical east midlands urban inns, shops and offices that range in date from the seventeenth to the twentieth century, but which are dominated by the more opulent structures built after the arrival of the East Coast Main railway line, in 1850. It's a shame that the first glimpse the modern railway passenger gets of the city centre is a series of brutalist concrete multi-storey car parks, festooned with swathes of wire netting to stop people jumping off them (sadly, several have done so).

Wherever you enter the concourse of Queensgate, you cannot fail to be drawn into the huge space before you. Like the car parks, the building has a steel and concrete core, but is faced with brick and pale stone panels: here, though, the lines are broken and unexpected, giving texture to the whole; glass, likewise, is used imaginatively. The combination of concrete and glass can be brutal, but certainly not here. Even if you don't like shops and shopping, the pull of Queensgate is magnetic: in this respect, it puts me in mind of the great Norman cathedral, just across the city centre. The main body of the shopping centre is a vast three-storey, extended hall-like space with a prominent raised walkway at first-floor level. High above your head is a light ceiling, which echoes the paleness of the daytime sky. When you are standing in the centre of the mall and look in either direction, the space appears to continue around corners. In fact, these are dead-ends, because in plan the layout of Queensgate resembles a straightened letter S, with foreshortened ends. But the effect is extraordinary: after many years of regular visits, I still think Queensgate is far larger than its actual size.

Queensgate is the culmination of the city's New Town, as planned and executed by the Peterborough Development Corporation. The New Towns were created after the war, mostly to remove 'overspill' population from Greater London. The idea was not to create entirely new communities, but to expand existing centres of population. Their progress was managed by newly appointed Development Corporations. The first wave of New Town Corporations included Stevenage (1946) and Harlow (1947); Milton Keynes and Peterborough (both 1967) were in the second.[1] I became involved with Peterborough in 1970, as part of the Development Corporation's efforts to deal with archaeological sites threatened by the New Town. One of the people I negotiated with was the Chief Planning Officer, David Bath, who would invariably be buried in work on Queensgate; its success, and that of the New Town more generally, mattered to him and the other senior staff personally. It was also quite apparent that he valued what it represented – it was to be a new focus for the city centre and not just a 'retail opportunity'.

In Peterborough, as in many New Towns, traffic was routed around the ancient city centre, leaving it free for pedestrians. The bypass also served as a boundary between areas of housing and industry on the city's periphery. I can remember a number of terrible early post-war buildings being removed from the city centre (a process which still continues). The result has been to greatly improve views of the Town Church, St John the Baptist, which now dominates Cathedral Square. This church, a fine example of fifteenth-century Perpendicular, was built with money from the wool trade when the town was sometimes known as Guildenburgh: city of gold. The pedestrianized Cathedral Square lies at the heart of the civic city centre. Its principal feature is the superb limestone Old Guildhall, built in 1671 with an open ground floor composed of columns and a roof of the distinctive local Collyweston limestone slate. This building symbolizes the increasing assertiveness of civic society in the early modern age; that symbolism is also apparent in the Guildhall's confrontational positioning, which is directly opposite the heavy, fortress-like medieval Cathedral Gatehouse.

The coming of the East Coast Main railway line had a major effect on the city, which by the mid-nineteenth century no longer enjoyed its earlier prosperity, following the decline of the wool trade in post-medieval times. The Great Northern Railway company's original intention was to route the line more to the west, through Stamford, but the powerful landowners there, the Cecils of Burghley House, would have none of it. So it went via Peterborough, which soon became an important railway junction, where lines to Lincolnshire, East Anglia and the midlands joined the main line to Scotland. Although the original station, devastated by dry rot, had to be demolished in 1976, some of the surviving contemporary buildings give a sense of the new dynamism that the railway brought to the city: the fine Georgian-style Great Northern Hotel and long, low brick-built warehouses that have now been converted into shops and offices.

Yet the railways also brought destruction. Almost every notable structure in the city – bar the Cathedral, the Town Church and the Old Guildhall – was demolished in the frenzy of new construction that accompanied the railways. But even that prosperity proved finite. When I first got to know Peterborough in the early 1970s, there seemed something under-confident about the place: always self-consciously aware that it was being compared unfavourably with neighbouring 'picturesque' Stamford. It was hardly a fair comparison, given that Peterborough's non-ecclesiastical early buildings were often made of timber, undressed limestone or soft local brick.

The west front of the Cathedral is, for me, unrivalled, especially seen in the low sunlight of late afternoon or evening.[2] The original Norman front to what was then the Benedictine Abbey now lies partly hidden behind three huge arches, which are held together, physically and visually, by a massive central porch. Work on the extended west front began in 1177, when an Abbot Benedict visited the city after a visit to Canterbury Cathedral, following the murder there of St Thomas Becket, four days after Christmas in 1170. The extension of the west front introduced to Peterborough the new, more ornate pointed-arch style of Gothic architecture. The building of the central porch, which was needed to prevent the great arches from spreading, finished these improvements and marked the Abbey's dedication, in the year 1238.

Many of the monastic buildings still survive, especially around the west front in Minster Yard. If you head round the Abbey's south side – moving through the space left by the long-dismantled old cloisters – to the east end, you find the exterior of the late Gothic Retrochoir or New Building. On the northern side of the Cathedral, meanwhile, you can see as fine a display of early Norman round arches as anywhere else in England. Inside, the panel-painted nave ceiling, dating to the second half of the thirteenth century, is the finest in Europe north of the Alps. It is arranged in carefully framed diamonds and triangles and features scenes involving various saints, kings and bishops. But it echoes and enhances the architecture beneath wonderfully. We know that nave ceilings were sometimes painted, but wooden roofs rot and often need repair and replacement, so all others (including that at Canterbury) have been lost. The last threat to Peterborough's nave ceiling was a terrible fire, in 2001. The conservation project that followed was truly heroic.

The round-arched stonework of the Cathedral's early Norman interior is superbly preserved and had required remarkably little Victorian restoration, with the exception of the area around the central tower. But the less known jewel-in-the-crown is the Retrochoir, which was added to the early Norman apsidal east end in the early sixteenth century. It is executed in the Late Gothic style of architecture that is unique to Britain and which features delicate fan-vaulting, slender arches and large windows. The Retrochoir is perhaps the least well-known example of this architectural style, which is better known at King's College Chapel, Cambridge and St George's Chapel, Windsor Castle – and it is every bit as good. The project to build the Retrochoir was undertaken by the penultimate Abbot, John Kirkton, or Kirton (1496–1528) and the highly decorated fan-vaulting was designed by John Wastell, before he began work on the ceiling of King's College Chapel.

Unlike many other great churches, the architecture of Peterborough Cathedral has not been obscured by too many monuments, but Catherine of Aragon and Mary Queen of Scots (who was executed at nearby Fotheringhay Castle) are commemorated in the Sanctuary with simple, yet moving, monuments. Despite these royal connections, Peterborough Cathedral remains one of the most under-appreciated of Britain's medieval churches. Of the Fens' three great cathedrals, Peterborough tends to be bypassed in favour of Ely and Lincoln, but in my opinion this is a huge mistake. It is a magnificent and remarkably complete building that symbolizes the region's prosperity in the Middle Ages and remains as fine an example of early medieval architecture as any cathedral in Britain.

24

King's Cross and St Pancras

Where History Matters

My earliest memory of King's Cross Station was as an eight- or nine-year-old, walking with my mother to the front of the train that was to take us to Hitchin. This was in the mid-1950s, and I had just been given my first model railway set. I was now obsessed with everything to do with tracks, signal boxes, stations and, of course, steam engines. Diesel and electric trains didn't get a look-in. My mother was both attractive and blessed with the charm of the Irish: unsurprisingly, the nice engine driver and fireman agreed to take me into their cab for the journey. The fireman sat me on some coal in the tender and, when we were under way, moved me to a small ledge-like seat at the side of the cab. The sparks and smoke flew as we slowly accelerated, with a deafening din, into the first of the tunnels immediately north of the station. In the confined space the noise, bouncing off the brickwork, was thunderous: I was thrilled and petrified. It was a mind-blowing experience. And it was happening to me! Ever since that time, King's Cross has occupied a special place in my heart.

The great stations that served the midlands and the north – Euston, St Pancras and King's Cross – were built along the northern fringes of mid-nineteenth-century London.[1] Euston was built first (in 1837); King's Cross second (1852); and St Pancras third (1868). The Euston Road, which was built a century earlier (when it was known as the New Road), still links them all together. It ran more or less along the edge of north London when Euston Station was opened, but three decades later, when St Pancras was constructed, the area was thoroughly urban. Such was the speed of development in the early railway age. And of course it made good sense to place buildings that needed as much space as railway termini on green- and brown-field sites, where land costs would be cheaper. Today London extends way beyond the Euston Road. As your train flashes or trundles through the northern suburbs, it's worth recalling that the rapidity of London's nineteenth- and early twentieth-century expansion was in large part paid for and funded by the explosion of trade and commerce that the railways themselves stimulated.[2]

The earliest of the major London termini, opened in 1837, was Euston. At first, what is now the West Coast Main Line only went as far as Birmingham, but soon the tracks were extended north, eventually to Glasgow. This magnificent station was largely destroyed in 1962 by a crass act of cultural vandalism that witnessed, among other horrors, the deliberate demolition of the famous Euston Arch: a gateway in the form of an ancient Greek temple that had become a symbol of the Railway Age. Its proposed destruction in the name of 'progress' caused such an outcry that it had to be sanctioned by the then Prime Minister, Harold Macmillan.

Euston began small, with just two platforms, before growing over the following decades. King's Cross, however, was very different. For my money it is the finest railway station in the world. And unlike poor damaged Euston, recent developments have, if anything, improved it.[3]

Designed for the Great Northern Railway by their architect, Lewis Cubitt, when King's Cross Station opened in 1852 it was the largest in Britain. Two broad iron and glass archways covered two long train sheds, or sheltered platforms. In those days the archways, which now accommodate platforms 1–8, housed just two: one for trains arriving, the other for trains departing. Sidings for carriages occupied the space in the middle. The magnificent southern front of the station is remarkably restrained and unadorned. It is built with yellowish-brown so-called 'London Stock Bricks', which were fashioned from local clays and which give so many elegant eighteenth- and nineteenth-century London terraces their distinctive colour. The station's front reflects and highlights the two arched train sheds behind, with plain, buttressed walls at each end, and two huge arches on either side of a fine central clock and bell tower. In style it is late Georgian or Italianate: simple, dignified, yet strong.

In 1972, British Rail built a single-storey booking hall in the concourse immediately outside the station. Although deliberately kept low, this unfortunate building completely disrupted the view of the station's superb façade. Happily, it was demolished during recent major renovations, which have proved to be practical, imaginative and highly successful, largely because they have treated Lewis Cubitt's building with the respect it deserves. Work began in 2005 on a major refurbishment of King's Cross, which included the construction of a huge, semi-circular domed concourse built against the south-western side of the station, behind the Great Northern Hotel. The breathtaking, net-like latticed steel roof, designed by John McAslan, is believed to be the largest single-span station structure in Europe, and in its scale, ambition and sympathy for its surroundings recalls Foster and Partners' superficially similar new atrium for the British Museum.

At the northern end of the concourse is what seems to be a permanent queue of young Harry Potter fans, waiting to be pictured by a luggage trolley for Platform 9¾, where, of course, the train departs for Hogwarts, via the Hogwarts Express. Beyond the queue is a flight of stairs leading up to my favourite destination at King's Cross: the Parcel Yard pub. As its name suggests, it is situated in the building where parcels were sorted. The brewers have done a splendid job in preserving the original decor, doors, flooring and windows, and the furniture is second-hand, slightly battered, and entirely appropriate. There are excellent views out over the two train sheds. The rooms and the building are all Grade I listed – the highest level of statutory protection. Every time I pass through the station, I raise a foaming pint of London Pride to the wisdom of that Listing decision. When the inevitable happens and trains are delayed, the Parcel Yard is always there to soothe one's growing anxiety.

Immediately west of King's Cross, just across Pancras Road, is St Pancras Station.[4] Today it is probably better known than its slightly earlier neighbour, largely due to the wonderfully over-the-top Victorian Midland Grand Hotel that forms its southern frontage, and which, with its profusion of turrets and fanciful patterned brickwork, provides a superbly Gothic contrast with the elegant restraint of King's Cross. Designed by the great Victorian architect George

Gilbert Scott, it took five years to build and was opened in 1873. Scott's Midland Hotel conceals a slightly earlier, more unusual, and structurally innovative building directly behind it – the train shed itself.

In designing St Pancras' train shed, its architect, William Henry Barlow, had to contend with a number of practical difficulties, mostly caused by the nearby presence of the early nineteenth-century Regent's Canal, built to link the Grand Union Canal to the London Docks on the Thames at Limehouse, and which included docks and basins in the then open country immediately north of King's Cross.

Barlow's chief headache was to come up with a scheme that allowed the rail line immediately north of St Pancras to cross the Regent's Canal while allowing sufficient headroom for barges. This meant that the track bed had to be raised to avoid steep gradients, resulting in a station at two levels, with the lower level, the undercroft, to be used for storage and for renting out. The first plan was for a three-arched train shed, but this had to be abandoned, as its foundation interfered with the layout of the lower level. In the end, Barlow opted for a single-span arched roof, which remains a breathtaking structure today. When the station opened for business in 1868, it was the largest single-span roof in the world.

In 1963, a plan for the rationalization of the nationalized British Railways network was drawn up by the engineer Dr Richard Beeching. Beeching recommended the closure of one of the lines that terminated at St Pancras, but his wider report had a profound influence on politicians and others who then controlled the railways. By the mid-1960s there was an increasing chorus of influential voices that were calling for the whole station's closure. At the same time it was discovered that the ironwork of the great roof arch was in urgent need of repair. Things were looking bad, but there was also a growing body of opinion that favoured conservation and which was now bitterly regretting the demolition of the Euston Arch in 1962. Chief among those that opposed the closure of St Pancras was the poet and broadcaster Sir John Betjeman, who was supported by the then recently formed Victorian Society. To cut a long story short, the conservation lobbyists eventually prevailed, and St Pancras was spared. It had been a close call.

St Pancras has been the subject of major redevelopment in the twenty-first century, culminating with the opening in 2007 of the high speed Eurostar service. Further development is happening in the area around both stations and to the north, towards and over the Regent's Canal and the pre-railway warehouses that surround it. It is now one of London's most important areas of growth, looking to the future while respecting the area's rich history. In St Pancras Station itself, homage is paid to Betjeman's influential role in a statue, slightly larger than life-size, by Martin Jennings. Holding on to his hat with one hand, Sir John looks up, marvelling as I do, every time I visit, at the great iron arch, high above our heads.

Notes

INTRODUCTION

1. Gwen Raverat, *Period Piece* (Faber & Faber, London, 1960), p. 159.
2. Tori Reeve, *Down House: The Home of Charles Darwin* (English Heritage, Swindon, 2009).
3. Ibid., p. 22.
4. Francis Pryor, *The Making of the British Landscape* (Penguin Books, London, 2011), p. 5.
5. Ibid., p. 641.
6. Francis Pryor, *Stonehenge: The Story of a Sacred Landscape* (Head of Zeus, London, 2016).

1. AFTER AGES OF ICE

1. J. G. D. Clark, *Excavations at Star Carr, an Early Mesolithic Site at Seamer, near Scarborough, Yorkshire* (Cambridge University Press, Cambridge, 1954).
2. I discuss this, with references, in Francis Pryor, *Home* (Penguin Books, London, 2015), p. 2.
3. Paul Mellars, Tim Schadla-Hall, Paul Lane and Maisie Taylor, 'Chapter 4: The Wooden Platform', in Paul Mellars and Petra Dark (eds.), *Star Carr in Context* (McDonald Institute Monographs, Cambridge, 1998), pp. 47–64.

2. ORKNEY ISLANDS

1. A substantial Neolithic settlement was recently discovered on the tiny, treeless Orkney island of Wyre. It has been radiocarbon dated to 3300–3100 BC. Daniel Lee and Antonia Thomas, 'Orkney's First

Farmers: Early Neolithic Settlement on Wyre', *Current Archaeology* no. 268 (July 2012), pp. 12–19.

2. Nick Card, 'The Ness of Brodgar: More than a Stone Circle', *British Archaeology* no. 128 (2013), pp. 14–21.

3. *Bathymetrical Survey of the Fresh-Water Lochs of Scotland, 1897–1909, Lochs of Orkney* (National Library of Scotland), p. 225.

4. Colin Richards and Carly Hilts, 'A Tale of Two Neolithics? Investigating the Evolution of House Societies in Orkney', *Current Archaeology* no. 318 (September 2016), pp. 30–35.

3. AVEBURY

1. Isobel Smith, *Windmill Hill and Avebury: Excavations by Alexander Keiller, 1925–1939* (Clarendon Press, Oxford, 1965).

2. Stuart Piggott, *The West Kennet Long Barrow, Excavations 1955–56*, Ministry of Works Archaeological Reports no. 4 (HMSO, London, 1962).

3. There are many examples of similar practices at Neolithic tombs, which I discuss (with references) in *Britain BC* (HarperCollins, London, 2003), chapter 8.

4. GREAT ORME COPPER MINES

1. C. J. Williams, *Great Orme Mines* (Northern Mine Research Society, Keighley, 1995).

2. I. Longworth, A. Herne, G. Varndell and S. Needham, *Excavations at Grimes Graves, Norfolk 1972–1976*, Fascicule 3 (British Museum Press, London, 1991).

5. A WHITE LINE IN TIME

1. Stephen Johnson, *Hadrian's Wall*, 2nd edn, (B. T. Batsford, London, 2004), p. 43.

2. Stephen Johnson, *Hadrian's Wall* (Batsford Ltd, London, 1989).

3. One of those trees, a sycamore, was voted Tree of the Year in 2016. See https://www.woodlandtrust.org.uk > Explore woods > Tree of the Year 2016.

4. David Breeze, *The Antonine Wall* (John Donald, Edinburgh, 2006).

5. The Vindolanda writing tablets are constantly providing new information. They are best seen online. Go to vindolanda.csad.ox.ac.uk.
6. This stretch of the Cumbrian coast is mapped in Johnson, *Hadrian's Wall*, p. 63.

6. SHIFTING SANDS OF TIME

1. The new circle is definitively described by David Robertson and colleagues in 'A Second Timber Circle, Trackways, and Coppicing at Holme-next-the-Sea Beach, Norfolk: Use of Salt- and Freshwater Marshes in the Bronze Age', *Proceedings of the Prehistoric Society* vol. 82 (2016), pp. 227–58.
2. I try never to be without A. D. Mills's *A Dictionary of English Place-Names* (Oxford University Press, London, 1991). Mills sees (p. 48) the name as ultimately deriving from *castrum* and Branodunum, which was probably a reduced Latin version of an earlier Celtic name.
3. See map in Francis Pryor, *Britain AD* (HarperCollins, London, 2004), p. 138.
4. I discuss the Anglo-Saxon population movements in two books: *Britain AD*, chapter 6; and (with more recent references and the influence of DNA studies) *Home* (Penguin Books, London, 2015), pp. 285–6.
5. A. F. Pearson, *The Roman Shore Forts: Coastal Defences of Southern Britain* (Tempus Books, Stroud, 2002).
6. John Hinchliffe, *Excavations at Brancaster, 1974 and 1977*, East Anglian Archaeology Report no. 23 (Norfolk Archaeological Unit, 1985).
7. *Brancaster Roman Fort (Branodunum), Norfolk: Archaeological Evaluation and Assessment of Results*, Wessex Archaeology (Salisbury, 2015).
8. Strange acoustic effects have been noted in many prehistoric monuments. I discuss this briefly in *Britain BC* (HarperCollins, London, 2003), p. 246 (with references). Recent research (released January 2017) into the acoustics of Stonehenge has noted similar distorting and enhancing effects (bbc.co.uk/news/uk/england/wiltshire > Stonehenge sounds recreated using virtual reality).

7. ARTHURIAN TINTAGEL

1. Francis Pryor, *Britain AD* (HarperCollins, London, 2004), pp. 22–35.
2. The tomb itself did not survive the Dissolution, but its site can still be seen at Glastonbury Abbey.

3. http://blog.english-heritage.org.uk/discoveries-excavations-tintagel-castle/.

8. A HAUNTING PLACE

1. Francis Pryor, *The Making of the British Landscape* (Penguin Books, London, 2011), p. 354, fig. 9.11.
2. Nikolaus Pevsner, *The Buildings of England. Yorkshire: The North Riding* (Penguin Books, Harmondsworth, 1996), p. 392.

9. THE SCOTTISH BORDERS

1. Archaeological excavation and research is continuing at Whithorn. Go to: www.whithorn.com.
2. For a map of sites which have produced early imported pottery see *Britain AD* (HarperCollins, London, 2004), p. 182.
3. The following draws heavily on Kitty Cruft, John Dunbar and Richard Fawcett, *The Buildings of Scotland: Borders* (Yale University Press, London, 2006).
4. *The Shorter Oxford English Dictionary: On Historical Principles* (Oxford University Press, Oxford, 1973), p. 199.

10. THE BOSTON STUMP

1. Simon Jenkins, *England's Thousand Best Churches* (Penguin Books, London, 1999), p. 363.
2. There is an excellent account of Boston's history in N. Pevsner, J. Harris and N. Antram, *The Buildings of England. Lincolnshire* (Penguin Books, London, 1989), pp. 153–61.
3. I discuss the building of the Stump in *Britain in the Middle Ages* (HarperCollins, London, 2006), pp. 166–8.

11. ROMNEY MARSH

1. Today the footpath along the Royal Military Canal is part of the much longer (163 miles) Saxon Shore Way from Gravesend, in Kent, to Hastings, in East Sussex.

2. Royal Commission on Historical Monuments (England), *Peterborough New Town: A Survey of Antiquities in the Areas of Development* (HMSO, London, 1969), pp. 40–44.

3. H. Godwin, *Fenland: Its Ancient Past and Uncertain Future* (Cambridge University Press, Cambridge, 1978), pp. 142–3.

4. There is an excellent short illustrated guide to the region: J. Campbell, *The Medieval Churches of Romney Marsh* (Romney Marsh Historic Churches Trust, Kent, 2012).

5. C. Greatorix, *The Shinewater Track: The Excavation of a Late Bronze Age Waterlogged Structure on the Willingdon Levels, near East-bourne, East Sussex, Report on Project Number 408* (Archaeology South-East, Hassocks, West Sussex, 1998); P. Clark (ed.), *The Dover Bronze Age Boat* (English Heritage, Swindon, 2004).

12. MEDIEVAL PRODUCTIVITY

1. There are others, but these have been substantially altered over time. I discuss Laxton and the Open Field System in *The Making of the British Landscape* (Penguin Books, London, 2011), pp. 296–314 (with references).

2. J. Claridge, 'The Role of Demesnes in the Trade of Agricultural Horses in Late Medieval England', *Agricultural History Review* vol. 65, Pt. 1 (2017), pp. 1–19.

3. Hugo J. P. La Poutré, 'Fertilization by Manure: A Manor Model Comparing English Demesne and Peasant Land, *c.* 1300', *Agricultural History Review* vol. 65, Pt. 1 (2017), pp. 20–48.

13. IRONBRIDGE

1. I discuss this at greater length in *The Birth of Modern Britain* (Harper-Collins, London, 2011), pp. 13–15.

2. Ibid., chapter 5.

3. I discuss Coalbrookdale in *The Making of the British Landscape* (Penguin Books, London, 2011), pp. 445–9 (with references).

15. THE QUIET REVOLUTION

1. I discuss the development of turnpikes in *The Making of the British Landscape* (Penguin Books, London, 2011), pp. 451–6; see also C. C.

Taylor, *The Cambridge Landscape* (Hodder and Stoughton, London, 1973), pp. 227–8.
2. For example: www.ourownhistory.com/walk_ermine_street.htm.
3. N. Pevsner, *The Buildings of England. Cambridgeshire* (Penguin Books, Harmondsworth, 1954), p. 246.

16. A BRIDGE WITHOUT SIDES

1. For Stonyhurst see N. Pevsner, *The Buildings of England. North Lancashire* (Penguin Books, Harmondsworth, 1969), pp. 239–45.
2. I have contacted the local authority, who have assured me that they are aware of the situation and are addressing it; https://historicengland.org.uk/advice/heritage-at-risk/search-register/list-entry/1684290.
3. Sadly, Pevsner got it wrong: it is neither medieval, nor a ruin: *North Lancashire*, p. 149.
4. The gate inn sign can now be seen on display in the museum at Elgood's Brewery, Wisbech, Cambridgeshire.

17. EDINBURGH NEW TOWN

1. I discuss Edinburgh New Town in *The Making of the British Landscape* (Penguin Books, London, 2011), pp. 554–6 (with references).
2. For a good account of the New Town go to: www.ewht.org.uk/learning/Athens/the-new-town-plan.

18. PERFECTION AT ROUSHAM

1. The best known of these genteel visitors must surely be Celia Fiennes, who has left us a fascinating account of her many expeditions to great country houses: C. Morris (ed.), *The Illustrated Journeys of Celia Fiennes c.1682–c.1712* (Webb and Bower, Exeter, 1982).
2. T. Williamson, *Polite Landscapes: Gardens and Society in Eighteenth-Century England* (Johns Hopkins University Press, Baltimore, 1995).
3. For a succinct account of Rousham see G. and S. Jellicoe, P. Goode and M. Lancaster, *The Oxford Companion to Gardens* (Oxford University Press, London, 1986), pp. 486–7.

4. For a beautifully illustrated (photos by Marcus Harpur) account of the gardens, see G. Plumptre, *The English Country House Garden* (Frances Lincoln, London, 2014), pp. 46–53.

19. THE CAUSEY ARCH, CO. DURHAM

1. I discuss the Causey Arch and tramways in *The Making of the British Landscape* (Penguin Books, London, 2011), pp. 513–17. See also, F. Aalen, *England's Landscape: The North East* (Collins, London, 2006), p. 94.

20. BIRKENHEAD PARK

1. I discuss Birkenhead Park in *The Making of the British Landscape* (Penguin Books, London, 2011), pp. 564–7.
2. Francis Pryor, *The Birth of Modern Britain* (HarperCollins, London, 2011), colour plate 28.

21. RISEHILL NAVVY CAMP

1. D. Brooke, 'Railway Navvies on the Pennines, 1841–71', *Journal of Transport History*, New Series, vol. 3, no. 1 (February, 1975), pp. 41–53.
2. M. Freeman, *Railways and the Victorian Imagination* (Yale University Press, London, 1999), p. 179.
3. I have discussed Risehill in *The Birth of Modern Britain*, pp. 106–18 (HarperCollins, London, 2011).
4. The *Time Team* excavation report is: *Risehill Tunnel Navvy Camp, Cumbria: Archaeological Evaluation and Assessment of Results*. Wessex Archaeology Report 68737 (Salisbury, 2008).

22. LORDSHIP REC, TOTTENHAM

1. For an excellent account of the expansion of London in the late nineteenth and twentieth centuries, see Trevor Rowley, *The English Landscape in the Twentieth Century* (Hambledon Continuum, London, 2006), chapter 4.

2. Francis Pryor, *The Making of the British Landscape* (Penguin Books, London, 2011), pp. 606–11 (with references).
3. By far the best explanation of flooding and water management is by Jeremy Purseglove, *Taming the Flood: Rivers, Wetlands and the Centuries-Old Battle Against Flooding*, 2nd edn (William Collins, London, 2015).
4. www.haringey.gov.uk > lordship-recreation-ground.
5. http://tottenham-summerhillroad.com/lordshiprecground.htm.
6. https://lordshiprec.org.uk.
7. https://en.m.wikipedia.org/wiki/Broadwater_Farm.

23. QUEENSGATE SHOPPING CENTRE, PETERBOROUGH

1. I discuss New Towns in *The Making of the British Landscape* (Penguin Books, London, 2011), pp. 620–22 (with references).
2. For more on Peterborough town and cathedral see C. O'Brien and N. Pevsner, *The Buildings of England. Bedfordshire, Huntingdonshire and Peterborough* (Yale University Press, London, 2014), pp. 582–658. See also, S. Jenkins, *England's Cathedrals* (Little, Brown, London, 2016), pp. 194–9. Recently the cathedral has been given the book it deserves: J. Foyle, *Peterborough Cathedral: A Glimpse of Heaven* (ACC Publishing Group, Woodbridge, Suffolk, 2017).

24. KING'S CROSS AND ST PANCRAS

1. I discuss the early history of London's railways in *The Making of the British Landscape* (Penguin Books, London, 2011), pp. 522–4.
2. For a superb discussion of the Railway Age and its culture, see M. Freeman, *Railways and the Victorian Imagination* (Yale University Press, London, 1999).
3. https://www.kingscross.co.uk/kings-cross-station; https://en.m.wikipedia.org/wiki/London_King's_Cross_railway_station.
4. https://en.wikipedia.org/wiki/St_Pancras_railway_station.

Illustration Credits

ALLEN LANE
an imprint of
PENGUIN BOOKS

Also Published

Sunil Amrith, *Unruly Waters: How Mountain Rivers and Monsoons Have Shaped South Asia's History*

Christopher Harding, *Japan Story: In Search of a Nation, 1850 to the Present*

Timothy Day, *I Saw Eternity the Other Night: King's College, Cambridge, and an English Singing Style*

Richard Abels, *Aethelred the Unready: The Failed King*

Eric Kaufmann, *Whiteshift: Populism, Immigration and the Future of White Majorities*

Alan Greenspan and Adrian Wooldridge, *Capitalism in America: A History*

Philip Hensher, *The Penguin Book of the Contemporary British Short Story*

Paul Collier, *The Future of Capitalism: Facing the New Anxieties*

Andrew Roberts, *Churchill: Walking With Destiny*

Tim Flannery, *Europe: A Natural History*

T. M. Devine, *The Scottish Clearances: A History of the Dispossessed, 1600-1900*

Robert Plomin, *Blueprint: How DNA Makes Us Who We Are*

Michael Lewis, *The Fifth Risk: Undoing Democracy*

Diarmaid MacCulloch, *Thomas Cromwell: A Life*

Ramachandra Guha, *Gandhi: 1914-1948*

Slavoj Žižek, *Like a Thief in Broad Daylight: Power in the Era of Post-Humanity*

Neil MacGregor, *Living with the Gods: On Beliefs and Peoples*

Peter Biskind, *The Sky is Falling: How Vampires, Zombies, Androids and Superheroes Made America Great for Extremism*

Robert Skidelsky, *Money and Government: A Challenge to Mainstream Economics*

Helen Parr, *Our Boys: The Story of a Paratrooper*

David Gilmour, *The British in India: Three Centuries of Ambition and Experience*

Jonathan Haidt and Greg Lukianoff, *The Coddling of the American Mind: How Good Intentions and Bad Ideas are Setting up a Generation for Failure*

Ian Kershaw, *Roller-Coaster: Europe, 1950-2017*

Adam Tooze, *Crashed: How a Decade of Financial Crises Changed the World*

Edmund King, *Henry I: The Father of His People*

Lilia M. Schwarcz and Heloisa M. Starling, *Brazil: A Biography*

Jesse Norman, *Adam Smith: What He Thought, and Why it Matters*

Philip Augur, *The Bank that Lived a Little: Barclays in the Age of the Very Free Market*

Christopher Andrew, *The Secret World: A History of Intelligence*

David Edgerton, *The Rise and Fall of the British Nation: A Twentieth-Century History*

Julian Jackson, *A Certain Idea of France: The Life of Charles de Gaulle*

Owen Hatherley, *Trans-Europe Express*

Richard Wilkinson and Kate Pickett, *The Inner Level: How More Equal Societies Reduce Stress, Restore Sanity and Improve Everyone's Wellbeing*

Paul Kildea, *Chopin's Piano: A Journey Through Romanticism*

Seymour M. Hersh, *Reporter: A Memoir*

Michael Pollan, *How to Change Your Mind: The New Science of Psychedelics*

David Christian, *Origin Story: A Big History of Everything*

Judea Pearl and Dana Mackenzie, *The Book of Why: The New Science of Cause and Effect*

David Graeber, *Bullshit Jobs: A Theory*

Serhii Plokhy, *Chernobyl: History of a Tragedy*

Michael McFaul, *From Cold War to Hot Peace: The Inside Story of Russia and America*

Paul Broks, *The Darker the Night, the Brighter the Stars: A Neuropsychologist's Odyssey*

Lawrence Wright, *God Save Texas: A Journey into the Future of America*

John Gray, *Seven Types of Atheism*

Carlo Rovelli, *The Order of Time*

Mariana Mazzucato, *The Value of Everything: Making and Taking in the Global Economy*

Richard Vinen, *The Long '68: Radical Protest and Its Enemies*

Kishore Mahbubani, *Has the West Lost It?: A Provocation*

John Lewis Gaddis, *On Grand Strategy*

Richard Overy, *The Birth of the RAF, 1918: The World's First Air Force*

Francis Pryor, *Paths to the Past: Encounters with Britain's Hidden Landscapes*

Helen Castor, *Elizabeth I: A Study in Insecurity*

Ken Robinson and Lou Aronica, *You, Your Child and School*

Leonard Mlodinow, *Elastic: Flexible Thinking in a Constantly Changing World*

Nick Chater, *The Mind is Flat: The Illusion of Mental Depth and The Improvised Mind*

Michio Kaku, *The Future of Humanity: Terraforming Mars, Interstellar Travel, Immortality, and Our Destiny Beyond*

Thomas Asbridge, *Richard I: The Crusader King*

Richard Sennett, *Building and Dwelling: Ethics for the City*

Nassim Nicholas Taleb, *Skin in the Game: Hidden Asymmetries in Daily Life*

Steven Pinker, *Enlightenment Now: The Case for Reason, Science, Humanism and Progress*

Steve Coll, *Directorate S: The C.I.A. and America's Secret Wars in Afghanistan, 2001 - 2006*

Jordan B. Peterson, *12 Rules for Life: An Antidote to Chaos*

Bruno Maçães, *The Dawn of Eurasia: On the Trail of the New World Order*

Brock Bastian, *The Other Side of Happiness: Embracing a More Fearless Approach to Living*

Ryan Lavelle, *Cnut: The North Sea King*

Tim Blanning, *George I: The Lucky King*

Thomas Cogswell, *James I: The Phoenix King*

Pete Souza, *Obama, An Intimate Portrait: The Historic Presidency in Photographs*

Robert Dallek, *Franklin D. Roosevelt: A Political Life*

Norman Davies, *Beneath Another Sky: A Global Journey into History*

Ian Black, *Enemies and Neighbours: Arabs and Jews in Palestine and Israel, 1917-2017*

Martin Goodman, *A History of Judaism*

Shami Chakrabarti, *Of Women: In the 21st Century*

Stephen Kotkin, *Stalin, Vol. II: Waiting for Hitler, 1928-1941*

Lindsey Fitzharris, *The Butchering Art: Joseph Lister's Quest to Transform the Grisly World of Victorian Medicine*

Serhii Plokhy, *Lost Kingdom: A History of Russian Nationalism from Ivan the Great to Vladimir Putin*

Mark Mazower, *What You Did Not Tell: A Russian Past and the Journey Home*

Lawrence Freedman, *The Future of War: A History*

Niall Ferguson, *The Square and the Tower: Networks, Hierarchies and the Struggle for Global Power*

Matthew Walker, *Why We Sleep: The New Science of Sleep and Dreams*

Edward O. Wilson, *The Origins of Creativity*

John Bradshaw, *The Animals Among Us: The New Science of Anthropology*

David Cannadine, *Victorious Century: The United Kingdom, 1800-1906*

Leonard Susskind and Art Friedman, *Special Relativity and Classical Field Theory*

Maria Alyokhina, *Riot Days*

Oona A. Hathaway and Scott J. Shapiro, *The Internationalists: And Their Plan to Outlaw War*

Chris Renwick, *Bread for All: The Origins of the Welfare State*

Anne Applebaum, *Red Famine: Stalin's War on Ukraine*

Richard McGregor, *Asia's Reckoning: The Struggle for Global Dominance*

Chris Kraus, *After Kathy Acker: A Biography*

Clair Wills, *Lovers and Strangers: An Immigrant History of Post-War Britain*

Odd Arne Westad, *The Cold War: A World History*

Max Tegmark, *Life 3.0: Being Human in the Age of Artificial Intelligence*

Jonathan Losos, *Improbable Destinies: How Predictable is Evolution?*

Chris D. Thomas, *Inheritors of the Earth: How Nature Is Thriving in an Age of Extinction*

Chris Patten, *First Confession: A Sort of Memoir*

James Delbourgo, *Collecting the World: The Life and Curiosity of Hans Sloane*

Naomi Klein, *No Is Not Enough: Defeating the New Shock Politics*

Ulrich Raulff, *Farewell to the Horse: The Final Century of Our Relationship*

Slavoj Žižek, *The Courage of Hopelessness: Chronicles of a Year of Acting Dangerously*

Patricia Lockwood, *Priestdaddy: A Memoir*

Ian Johnson, *The Souls of China: The Return of Religion After Mao*

Stephen Alford, *London's Triumph: Merchant Adventurers and the Tudor City*

Hugo Mercier and Dan Sperber, *The Enigma of Reason: A New Theory of Human Understanding*

Stuart Hall, *Familiar Stranger: A Life Between Two Islands*

Allen Ginsberg, *The Best Minds of My Generation: A Literary History of the Beats*

Sayeeda Warsi, *The Enemy Within: A Tale of Muslim Britain*

Alexander Betts and Paul Collier, *Refuge: Transforming a Broken Refugee System*

Robert Bickers, *Out of China: How the Chinese Ended the Era of Western Domination*

Erica Benner, *Be Like the Fox: Machiavelli's Lifelong Quest for Freedom*

William D. Cohan, *Why Wall Street Matters*

David Horspool, *Oliver Cromwell: The Protector*

Daniel C. Dennett, *From Bacteria to Bach and Back: The Evolution of Minds*

Derek Thompson, *Hit Makers: How Things Become Popular*

Harriet Harman, *A Woman's Work*

Wendell Berry, *The World-Ending Fire: The Essential Wendell Berry*

Daniel Levin, *Nothing but a Circus: Misadventures among the Powerful*

Stephen Church, *Henry III: A Simple and God-Fearing King*

Pankaj Mishra, *Age of Anger: A History of the Present*

Graeme Wood, *The Way of the Strangers: Encounters with the Islamic State*

Michael Lewis, *The Undoing Project: A Friendship that Changed the World*

John Romer, *A History of Ancient Egypt, Volume 2: From the Great Pyramid to the Fall of the Middle Kingdom*

Andy King, *Edward I: A New King Arthur?*

Thomas L. Friedman, *Thank You for Being Late: An Optimist's Guide to Thriving in the Age of Accelerations*

John Edwards, *Mary I: The Daughter of Time*

Grayson Perry, *The Descent of Man*

Deyan Sudjic, *The Language of Cities*

Norman Ohler, *Blitzed: Drugs in Nazi Germany*

Carlo Rovelli, *Reality Is Not What It Seems: The Journey to Quantum Gravity*

Catherine Merridale, *Lenin on the Train*

Susan Greenfield, *A Day in the Life of the Brain: The Neuroscience of Consciousness from Dawn Till Dusk*

Christopher Given-Wilson, *Edward II: The Terrors of Kingship*

Emma Jane Kirby, *The Optician of Lampedusa*

Minoo Dinshaw, *Outlandish Knight: The Byzantine Life of Steven Runciman*

Candice Millard, *Hero of the Empire: The Making of Winston Churchill*

Christopher de Hamel, *Meetings with Remarkable Manuscripts*

Brian Cox and Jeff Forshaw, *Universal: A Guide to the Cosmos*

Ryan Avent, *The Wealth of Humans: Work and Its Absence in the Twenty-first Century*

Jodie Archer and Matthew L. Jockers, *The Bestseller Code*

Cathy O'Neil, *Weapons of Math Destruction: How Big Data Increases Inequality and Threatens Democracy*

Peter Wadhams, *A Farewell to Ice: A Report from the Arctic*

Richard J. Evans, *The Pursuit of Power: Europe, 1815-1914*

Anthony Gottlieb, *The Dream of Enlightenment: The Rise of Modern Philosophy*

Marc Morris, *William I: England's Conqueror*

Gareth Stedman Jones, *Karl Marx: Greatness and Illusion*

J.C.H. King, *Blood and Land: The Story of Native North America*

Robert Gerwarth, *The Vanquished: Why the First World War Failed to End, 1917-1923*

Joseph Stiglitz, *The Euro: And Its Threat to Europe*

John Bradshaw and Sarah Ellis, *The Trainable Cat: How to Make Life Happier for You and Your Cat*

A J Pollard, *Edward IV: The Summer King*

Erri de Luca, *The Day Before Happiness*

Diarmaid MacCulloch, *All Things Made New: Writings on the Reformation*

Daniel Beer, *The House of the Dead: Siberian Exile Under the Tsars*

Tom Holland, *Athelstan: The Making of England*

Christopher Goscha, *The Penguin History of Modern Vietnam*

Mark Singer, *Trump and Me*

Roger Scruton, *The Ring of Truth: The Wisdom of Wagner's Ring of the Nibelung*

Ruchir Sharma, *The Rise and Fall of Nations: Ten Rules of Change in the Post-Crisis World*

Jonathan Sumption, *Edward III: A Heroic Failure*

Daniel Todman, *Britain's War: Into Battle, 1937-1941*

Dacher Keltner, *The Power Paradox: How We Gain and Lose Influence*

Tom Gash, *Criminal: The Truth About Why People Do Bad Things*

Brendan Simms, *Britain's Europe: A Thousand Years of Conflict and Cooperation*

Slavoj Žižek, *Against the Double Blackmail: Refugees, Terror, and Other Troubles with the Neighbours*

Lynsey Hanley, *Respectable: The Experience of Class*

Piers Brendon, *Edward VIII: The Uncrowned King*

Matthew Desmond, *Evicted: Poverty and Profit in the American City*

T.M. Devine, *Independence or Union: Scotland's Past and Scotland's Present*

Seamus Murphy, *The Republic*

Jerry Brotton, *This Orient Isle: Elizabethan England and the Islamic World*

Srinath Raghavan, *India's War: The Making of Modern South Asia, 1939-1945*

Clare Jackson, *Charles II: The Star King*

Nandan Nilekani and Viral Shah, *Rebooting India: Realizing a Billion Aspirations*

Sunil Khilnani, *Incarnations: India in 50 Lives*

Helen Pearson, *The Life Project: The Extraordinary Story of Our Ordinary Lives*

Ben Ratliff, *Every Song Ever: Twenty Ways to Listen to Music Now*

Richard Davenport-Hines, *Edward VII: The Cosmopolitan King*

Peter H. Wilson, *The Holy Roman Empire: A Thousand Years of Europe's History*

Todd Rose, *The End of Average: How to Succeed in a World that Values Sameness*

Frank Trentmann, *Empire of Things: How We Became a World of Consumers, from the Fifteenth Century to the Twenty-First*

Laura Ashe, *Richard II: A Brittle Glory*

John Donvan and Caren Zucker, *In a Different Key: The Story of Autism*

Jack Shenker, *The Egyptians: A Radical Story*

Tim Judah, *In Wartime: Stories from Ukraine*

Serhii Plokhy, *The Gates of Europe: A History of Ukraine*

Robin Lane Fox, *Augustine: Conversions and Confessions*

Peter Hennessy and James Jinks, *The Silent Deep: The Royal Navy Submarine Service Since 1945*

Sean McMeekin, *The Ottoman Endgame: War, Revolution and the Making of the Modern Middle East, 1908–1923*

Charles Moore, *Margaret Thatcher: The Authorized Biography, Volume Two: Everything She Wants*

Dominic Sandbrook, *The Great British Dream Factory: The Strange History of Our National Imagination*

Larissa MacFarquhar, *Strangers Drowning: Voyages to the Brink of Moral Extremity*

Niall Ferguson, *Kissinger: 1923-1968: The Idealist*

Carlo Rovelli, *Seven Brief Lessons on Physics*

Tim Blanning, *Frederick the Great: King of Prussia*

Ian Kershaw, *To Hell and Back: Europe, 1914–1949*

Pedro Domingos, *The Master Algorithm: How the Quest for the Ultimate Learning Machine Will Remake Our World*

David Wootton, *The Invention of Science: A New History of the Scientific Revolution*

Christopher Tyerman, *How to Plan a Crusade: Reason and Religious War in the Middle Ages*

Andy Beckett, *Promised You A Miracle: UK 80–82*

Carl Watkins, *Stephen: The Reign of Anarchy*

Anne Curry, *Henry V: From Playboy Prince to Warrior King*

John Gillingham, *William II: The Red King*

Roger Knight, *William IV: A King at Sea*